THE
RACE
FOR
MARS

'Mars. Planet of deserts. A vast wilderness of drifting sands, broken rock scoured by the wind, mist-filled valleys, lofty volcanoes, and snowy polar ice fields. A world whose peach skies are streaked by white wisps of cirrus cloud and thunderous dust storms. There are multitudes of reasons for going to Mars. It is the most Earth-like of all the planets in our solar system. Its weather, geological processes, and ancient history can provide clues about the natural forces at work on Earth, which will have a direct bearing on our everyday life.'

Astronomy,
June, 1985.

THE RACE FOR MARS

PETER HAINING

A COMET BOOK

A Comet Book
Published in 1986
by the Paperback Division of
W. H. Allen & Co Plc
44 Hill Street, London W1X 8LB

Typeset by Phoenix Photosetting, Chatham
Printed and bound in Great Britain by
Mackays of Chatham Ltd, Kent

ISBN 0-86379-044-5

For
My son, Richard
who will see the first Martians

CONTENTS

FOREWORD

THIS BOOK IS not science fiction; it is science fact. The race with which it is concerned is not an imaginary one, but has been declared in words like these by highly placed officials of the two competing superpowers:

'The space shuttle will enable us to operate routinely 22,000 miles above the Earth, and from there realize the dream of mounting a manned expedition to Mars.'

James M. Beggs,
NASA Administrator, 1984

'The Soviet Union could be ready to send a manned mission to Mars in 10 to 15 years and such a mission is fully within reach of Soviet technology. The discovery of evidence of life on Mars by unmanned spacecraft could even further spur a Soviet manned mission.'

Konstantin Feotistov,
USSR space scientist, 1985

What follows is the story of this race, how it began, how it has gathered momentum, and how it now at last stands on the threshold of being won. . . .

ONE
Mission – The Red Planet

I HAVE BEEN TO Mars. I have journeyed by interplanetary spaceship to the red planet and there roamed the vastness of its rocky deserts and explored the mysterious terrain of its valleys and mountains.

All in my imagination, of course, in company with Edgar Rice Burroughs, Arthur C. Clarke, Ray Bradbury and those other Martian chroniclers whose visions of this most intriguing world have made it almost real enough to experience. Soon, though, perhaps sooner than most of us know, we *will* go to Mars. Quite literally – the dream of centuries will become a reality.

Probably within the life-span of many people alive today, and *certainly* during that of our children, members of the human race will stand on the surface of our nearest planetary neighbour in space. And there they will at last be able to solve the many intriguing puzzles about this strange world which has fascinated us for centuries.

Already, of course, machines from Earth have circled Mars, and some have even landed. Their observations and reports, their superb eyewitness photographs and endless stream of data, have only served to increase our enthusiasm for the world which has been described as being the most similar to Earth anywhere in the solar system. Now man himself is poised to follow these pioneer vehicles, and before the end of the twentieth century there seems every chance that members of our species will finally step onto another planet and become, in effect, Martians.

But *who* will be first? For just as the Americans and Russians – the world leaders in space travel – competed with all their massive technological wizardry to put the first man on the moon, so the evidence is available to show there is a new race on between the superpowers. A race with a far richer prize at stake than that barren satellite which circles our Earth. As I shall show, there is more – far more – than just the kudos of being first to Mars: for this is a world rich in possibilities and probabilities. A world loaded with mineral deposits and the conditions which *could* support colonization and agriculture. A world, in truth, where man might settle and make a new beginning in the life of the universe.

But this is not just my opinion. Listen to these words of Thomas O. Paine who was the administrator of NASA during the first Apollo moon landing in 1969.

'The resource-rich planet Mars is man's next challenge in space,' he says. 'Yuri Gagarin blazed the trail to orbit; Neil Armstrong pioneered mankind's 'giant leap' to the moon. The men and women who will open the high road to Mars are alive today . . . and the technologies critical to the establishment and support of permanent Martian bases are already well developed.

'Apollo astronauts demonstrated effective techniques for exploring other worlds as far back as 15 years ago,' Mr Paine continues. 'Martian landings will prove easier because such techniques as aerobraking can be employed in the Martian atmosphere which, of course, the moon lacked. Both the American shuttle and the Soviet Union's giant new Series G launch vehicle, with its estimated cargo capacity of 200 tons, can easily reach low Earth orbit. Within a decade both the United States and the Soviet Union will have spaceports capable of assembling, loading, fuelling and launching deep-space transports to Mars.'

Such confidence is not based on dreams and wishful thinking. Both America and Russia have already demonstrated superb skill at putting men and machines into space, enabling them to live and function for increasingly lengthy periods of time, and there in the vastness of space carry out the whole range of activities a journey and landing on Mars will require – not the least of these, returning home thereafter!

8

Mars has been called 'the most user-friendly planet in the solar system' and certainly we on Earth have looked upon it with a mixture of interest and awe since the earliest times. The basic facts about the unmistakable ochre-red world which this year (1986) outshines all the other planets in our heavens except Venus, are now well-established and worth clarifying before we go any further.

Mars is the fourth planet from the sun, after Mercury, Venus and Earth. Like those planets it has a mostly solid surface, in stark contrast with the outer planets which are composed largely of gas. It moves around the sun at a distance of 141 million miles in comparison to the Earth's distance of 93 million miles. During its orbits, it comes as close to the Earth as 40 million miles (about once every 780 days) and as far distant as 250 million miles – facts which account for the variable way in which it can be seen by telescopes or the naked eye.

The planet is just slightly more than one-half as large as the Earth, its diameter at the equator being 4,221 miles. Its mass, though, is only one-tenth that of Earth, which means its gravitational pull is far less.

The Martian day is just over 24 hours long, but a year there lasts 687 days or 1.9 Earth years. This is caused by the fact that the rotation period of the planet is slightly more than half an hour longer than ours. The axial tilt of 25 degrees is, though, much the same as ours, which explains why Mars enjoys seasons *rather* like our own though they are, of course, much longer in duration. I say only rather like ours, because Mars' eccentric orbit around the sun causes these seasons to be of unequal lengths and intensities in each hemisphere. The south has short, hot summers and long, cold winters, while the seasons in the north are less extreme.

The atmosphere on Mars is very thin, with a barometric pressure of less than ten millibars, or approximately one one-hundredth of that on Earth. It consists almost solely of unbreathable carbon dioxide gas. For most of the Martian year, this atmosphere is relatively calm, with wind speeds rarely exceeding 11 miles per hour. However, during the 'summer' in the southern half of the planet, violent storms can develop with winds in excess of 110 mph. On occasions, dust storms have been so enormous that they cover the entire globe in a dusty veil!

Temperatures on Mars also have a wide daily and seasonal variation. At the poles, for example, during the winter, these may reach as low as −140 °C (−225 °F), whereas the noon-day temperatures at the equator can climb to 20 °C (68 °F). Actual surface temperatures measured by one of the American Viking lander spacecraft varied by as much as 75°F between pre-dawn and mid-afternoon measurements. These were taken during the summer period in the northern hemisphere. This range can be attributed to the very thin atmosphere which is poor at retaining the heat, and nights on Mars are always a great deal colder than anywhere on Earth.

Observers of the planet have rightly declared that it is a geologist's paradise. There are many features which, though they are also to be seen on Earth, are found here on a vast scale which is made all the more awesome because of Mars' size.

Reddish-ochre deserts strewn with rocks and boulders dominate the scene, and these have been compared to our Sahara Desert, though they are infinitely more chilly. There are also huge volcanoes which tower to heights almost three times that of Mount Everest, and there are signs that they have been active for a very long period of time – and may well continue to be so today.

There are great canyons across the surface, and huge dry channels which stand in mute testimony to what we believe were past gigantic floods. There are also smaller channels, indicative of slow erosion by running water in the same manner that terrestrial river valleys form.

These channels, in fact, present one of the puzzles that the first human visitors will look into, because liquid water cannot exist on the Martian surface under present conditions. It would either freeze or rapidly evaporate. The channels therefore suggest different climatic conditions in the past.

The huge canyons indicate faulting and landsliding on a grand scale, while layered deposits at the poles suggest sustained cyclic sedimentation. Although no liquid water is at the surface, considerable evidence exists for ground ice, below which liquid water could be present. Indeed, most impact craters on the Martian surface are surrounded by flow patterns that indicate that the material ejected had a mud-like consistency, as if the impact had penetrated the permafrost layer to excavate the water-rich material from below.

This much is still speculation – but there seems little doubt that the white deposits which form the planet's 'polar caps' (and which have been observed from Earth for two hundred years and more) look like ice and snow for a very simple reason – they are! Or to be exact, they are formed of compressed ice, both ordinary water ice and a layer of 'dry ice' or solid carbon dioxide. They, in fact, wax and wane with the seasons, and it has been suggested that the carbon dioxide may well 'migrate' from pole to pole between the winter season of each hemisphere.

A regular seasonal event in the polar regions is the formation of clouds of carbon dioxide ice particles during the 'autumn', as gas starts to condense out of the atmosphere onto the growing cap. So much of the atmosphere condenses out in this process that atmospheric pressure decreases more than 30 per cent from autumn to winter. Actually, this pressure decrease is slightly smaller in the northern hemisphere because the polar cap is smaller.

Other cloud activity on Mars is related to water in the atmosphere. Although these amounts are very small indeed, the atmosphere is close to saturation much of the time and a wide variety of water ice clouds can be seen.

In appearance, Mars is very asymmetric. Most of the southern hemisphere is densely cratered and superficially resembles the highlands on the moon. It is made up of old regions which are believed to reflect Mars' early history and may be as much as four billion years old. In contrast, the northern hemisphere consists mainly of plains and is relatively sparsely cratered, although it has many large volcanoes which, of course, have no lunar counterparts. There are three main regions of volcanoes: in the Hellas Basin, the Elysium region, and by far the most prominent, Tharsis, where there are several very large volcanoes that resemble terrestrial shield volcanoes such as those to be found in Hawaii. Everything about them, though, is vastly larger.

Tharsis is the most picturesque as well as the highest region on Mars – and, being plainly visible on the surface like a huge bump, is often referred to as 'the Tharsis bulge'. At its heart stand the four massive volcanoes of Arsia Mons, Pavonis Mons, Ascreaus Mons and the superlative Olympus Mons, all towering over sixteen miles high.

Just to the east of Tharsis is another instantly recognizable feature, the 1,500-mile-long series of interconnected canyons which make up the Valles Marineris. Huge beyond anything on Earth (they are at least three and a half times deeper than the Grand Canyon) they are aligned radial to the centre of the Tharsis bulge and appear to be related in some way to the fractures observed in the region.

The preservation of features on the surface of Mars that are clearly billions of years old indicates extremely low erosion rates. Small channels in the old cratered terrain are evidence of an early period of fluvial action, but survival of the old craters indicates that the period was short. For most of the planet's history it would seem, then, that wind has probably been the main erosive agent.

However, despite the giant dust storms I have mentioned, the wind clearly has not been very efficient in eroding the surface because so much old terrain survives. Most of the wind's action therefore probably involves reworking previously eroded debris.

One cannot, of course, close any discussion of the geophysics of Mars without mentioning the planet's two small moons, discovered in 1877 by the American astronomer, Asaph Hall, and suitably named Deimos (Terror) and Phobos (Fear), after the two attendants of the war god, Mars.

Though both will be discussed later, we should just note that Phobos, the inner and larger of the two, has a diameter of 15 miles and circles Mars every 7.6 hours at an altitude of 3,700 miles; while Deimos is just 7.5 miles across and circles the planet every 1.3 days at a height of 15,000 miles. Both are strikingly different: Phobos has sharply-defined craters and is criss-crossed by numerous fractures; Deimos is much smoother and with smoother craters. Why these two moons should be so irregular in shape and different in appearance is another mystery which has led to some interesting speculation, as we shall also see later.

Having looked at the basics of Martian geology, the most pressing question which remains is obviously this: is there *any* evidence of life on Mars? Does all that we have learned mean it is a completely sterile world – and always has been?

Not necessarily, is what cautious Mars-watchers say, like Britain's Patrick Moore:

'The results we have had so far from the space vehicles which landed on Mars have been inconclusive,' he says. 'But we *may* have tried the wrong experiments; we may be making wrong interpretations of the admittedly curious Martian chemistry; we may even have looked in the wrong places. But all in all, it now seems certain that if there is any living thing on Mars, it must be very primitive indeed.'

However, the redoubtable Mr Moore is quick to admit that things on Mars might well have been very different in the past.

'As we have seen, the planet has lofty volcanoes and there is evidence of drainage systems associated with them. There seems no reasonable doubt that the features which look like river beds really *are* river beds, and the polar caps contain a great deal of water ice. If running water once flowed on Mars, the atmosphere must have been much denser than it is today.

'So Mars may have lost its atmosphere fairly quickly. If this is so, then there may be periods when the atmosphere thickens up sufficiently for rain to fall and water to flow. We may therefore be seeing Mars at its very worst; and if there really are fertile periods, then there is a chance that life would have appeared whether or not it could survive when conditions worsened once more.'

Such an exciting thought makes a manned mission to Mars not only a challenge – but surely a necessity, too. For isn't mankind's most absorbing obsession to know whether we are truly alone in the vastness of the universe – or whether we might share it, even *have* shared it, with another life form?

Two other interested observers of Mars, as well as being

keen supporters of manned landings, are the science fiction writers, Ray Bradbury from America and Britain's Arthur C. Clarke, both of whom have done much to promote public interest through their excellent stories about the red planet. Ray it was who boldly declared years ago, 'If the moon was one large step for mankind, Mars is the next largest,' – so let us talk to him first.

Ray Bradbury (1906–) has written numerous tales over the years about Mars, but is perhaps best known for *The Martian Chronicles* (1950) widely acknowledged as a classic of SF, and to many of his admirers the finest of all his works. It is a remarkable book by any standards, presenting as it does the story of the attempts by humans to colonize Mars, and of their repeated, ambiguous meetings with Martians who have the ability to change their shapes. That the *Chronicles* is very close to Ray's heart is self-evident when you hear him confess, 'I have always looked on myself as some kind of Martian!'

He elaborates on this statement with typical gusto. 'My affinity for the planet is immense and prolonged and most affectionate-fine,' he says. 'I make the claim of being a Martian because we are all Aristotle's children, which is to say children of the universe. Not just Earth, or Mars, or this system, but the whole grand fireworks. And if we are interested in Mars at all, it is only because we wonder over our past and worry terribly about our possible future.'

In another of his stories about Mars, 'Dark They Were, and Golden-Eyed', Ray has attempted to outline what he sees as the future for mankind on the planet.

'In the story I told of a man and his family who helped colonize Mars,' he explains. 'They lived in its strange seasons, and then stayed on when everyone else went back to Earth, until the day finally arrived when they found the odd weather and peculiar temperatures had melted their flesh into new shapes, tinted their skin, and put flecks of gold into their now fantastic eyes. And so they moved up into the hills to live in the old ruins and became – Martians!'

This, says Ray, is what will happen to us on that far world. 'The ruins may not be there,' he declares, 'but if necessary we will *build* the ruins, and live in them and name ourselves as the transplanted Earthmen of my story did. And we will not be of Earth any more, but truly Martians – just as in the far future we will be moon creatures and then, God and Time willing, benevolent circumnavigators of an as yet unselected target-sun.'

There is much of the poet and dreamer in Ray Bradbury as you might guess, but his belief in mankind reaching and colonizing Mars is backed by knowledge as well as intuition. He has been a studious observer of his country's missions in space, and tells an interesting story of what happened when the first American spacecraft landed on Mars in 1976. He was, he says, somewhat tongue-in-cheek, half hoping the machine's cameras might have picked up the sight of a group of Martians waving signs which read: BRADBURY WAS RIGHT! But the experience was rewarding in quite another way.

'I was over at the NASA Jet Propulsion Laboratory celebrating with all these laughing, dancing, crying people who'd spent their lives on their dream,' he says. 'A lot of people had made fun of them, said we were never going to go to Mars. And, suddenly, there we were, with Viking, on Mars.

'The next morning they put me on television around the world on Telstar. And the interviewer said to me, "Mr Bradbury, you've been writing about life on Mars all of your life. Now the first photographs have come through, and there is no life on Mars. How do you feel about that?"

'My answer was, "Fool! There *is* life on Mars. And it is us!"

Interestingly, these words were echoed at much the same time by Arthur C. Clarke (1917–) when addressing a seminar in the USA. 'I will make one prediction,' he told his audience after a lengthy discussion on mankind's future plans in space. 'Whether or not there is life on Mars now, there will be by the end of this century!'

Arthur also shares Ray Bradbury's unbounded enthusiasm for Mars, but has brought a perhaps more practical approach to writing about manned landings. Since his early days as a fan of science fiction in the thirties and forties – followed by his

15

period as chairman of the British Interplanetary Society which actually proposed a feasible moon rocket much like the one which eventually carried men to the moon – Arthur has demonstrated a fascination with Mars in both his factual and fictional works. Like Ray Bradbury, he is internationally known for a book about the red planet, *The Sands of Mars*, written in 1951, and subsequently never out of print. He also offered convincing scenarios for rocket flights to Mars in two early non-fiction works, *The Exploration of Space* (1951) and *The Challenge of the Spaceships* (1959). Though somewhat dated today, these books are still remarkable examples of Clarke's uncanny powers of prophecy.

In them, Arthur has outlined the path which will lead man to Mars. First, the creation of space stations orbiting the Earth to be used as the starting points for these missions. Secondly, the development of techniques to enable men to spend long periods in space. And, thirdly, detailed observation of the planet itself to facilitate both landing and colonization.

In the late 1960s, Arthur was already speaking of landing on Mars as being feasible 'sometime in the eighties', and committing himself on the idea of life on Mars as well as the benefits of reaching the planet.

'Nothing we've discovered about Mars has ruled out the likelihood there *might* be some form of organic life there. We've known for some time there's no possibility of higher animal life forms similar to those on Earth. But there may be life forms quite different from anything we have here. . . .'

And the benefits? 'They will be largely scientific,' he said, 'and they will be enormous because any discovery on a new planet produces vast quantities of knowledge which is valuable as part of man's heritage and which inevitably has all sorts of unexpected repercussions and practical applications which can never be foreseen. I feel reasonably sure we'll be living on Mars some day and many people will eventually call it home and perhaps look down on Earth as a terrible place and be glad they're on Mars!'

Today, Arthur's backing for going to Mars is even stronger. Indeed, in 1985 he sent a special videotape message from his home in Sri Lanka to the American House of Representatives in Washington, where it was read into the Congressional Record. Entitled, 'A Martian Odyssey', it drew on a historical maritime analogy to make its forceful and pragmatic point.

'Only eight years from now,' Arthur said, 'it will be exactly half a millennium since three tiny ships sailed forth from Spain to change the history of our species. And three is about right for the smallest Mars expedition!'

The cost of such a mission, Arthur went on, would be 'less than that proposed merely for research into an anti-ICBM system' and would bring mankind the triumph of stepping once again on a completely new shore.

And in a dramatic appeal, he added, 'So is it absurdly optimistic to hope that by Columbus Day 1992, the United States and the Soviet Union will have emerged from their long winter of sterile confrontation? That would be none too soon to start talking seriously about mankind's next, and greatest, adventure.'

With these words, the old prophet of SF sowed a wholly new idea about putting mankind on Mars: that it might be achieved by nations – *jointly*. We shall be returning to this thought-provoking and challenging idea later in the book.

The reader will doubtless have appreciated from just these few pages what an exciting and fascinating prize Mars offers to the first men to land there. And though the Americans and Russians are more than aware of this fact, it would be wrong to imagine that this is a newly-formed belief on their respective parts, or that the competitive interest of the two nations has been sparked purely by the Americans' triumph on the moon. Men in both countries have yearned after the mysterious red planet for generations – in truth, since rocket flight was perceived as the first feasible method of interplanetary travel. The story of these men, their separate endeavours, and the dream which drove them all on, forms the fascinating, colourful and at times extraordinary background to *The Race For Mars*, which I have pieced together from both US and Soviet sources for the first time.

However, before we can even begin to study this part of our story and how it has brought us to the verge of a dream realized, it is necessary to turn back the pages of history and examine mankind's ageless fascination with Mars. The American space scientist Bruce Murray has said of this planet, with brilliant insight, 'We *want* Mars to be like the Earth. There is a very deep-seated desire to find another place where we can make another start, that somehow could be habitable. . . .'

□

Because Mars is a planet that is visible to the naked eye, it is not surprising to find that it was known to mankind long before the advent of the telescope. Perhaps equally predictably, because of its red colour, it was immediately associated with visions of fire, blood and war. The Chaldeans, for instance, called it Nergal, after their god of battle; while to the Persians it was 'Pahlavania Siphir' – the celestial warrior. The Greeks named the fiery red ball Ares, a derivative of their word 'to kill' or 'disaster', while the Romans, of course, gave it the name by which it is now familiar, after their god of war.

The early astronomers who first tried to distinguish any features on the planet were somewhat hampered in their observations by the then-prevalent belief that the Earth was the centre of the universe, and because Mars increased and decreased in brightness it must be just another of the 'wandering' bodies in the heavens. The first astronomer to establish that Mars actually moved around the sun and had different phases in its appearance was the Italian, Franciscus Fontana (1608–69), whose sketch made on 24 August 1638 is widely familiar. Although there has been much debate about the spot in the centre of the planet, experts are now agreed this was a fault in the telescope rather than any attempt to delineate a large feature.

The first man to show surface features as such on Mars was the remarkable Dutch physicist and astronomer, Christiaan Huygens (1629–93). He observed the planet on the night of 28 November 1656 and produced the first of several sketches showing a series of lines on the little world. Such was his excitement at this discovery that Huygens became one of the very earliest writers to speculate on the possibility of life on

Mars, as this extract from his posthumously published work, *Kosmotheros* (1798), clearly evidences:

'Mars has some parts darker than others, by the constant returns of which the days and nights have been found to be of about the same length as ours. But the inhabitants have no perceivable difference between summer and winter, the axis of the planet having very little inclination to the orbit. Our Earth must appear to them almost as Venus does to us, and by the help of a telescope will be found to have its wane, increase, and full like the moon. . . . Its light and heat is twice and sometimes three times less than ours, to which I suppose the constitution of its inhabitants is answerable.'

Another Italian astronomer, Giovanni Cassini (1625–1712), took the pioneer work of his fellow-countryman Fontana a whole step further in the spring of 1666 when his observations from Bologna, while Mars was in good view, resulted in the array of drawings shown here. Not only did they excite the world of astronomy, but helped develop an increasing public awareness about the planet Mars.

Cassini's nephew, Giacomo Maraldi (1646–1729), was just one of many to be caught up in this excitement, and from 1672 he devoted the most intense study to the planet. By the turn of the century he was convinced of two things. Firstly, that what could be seen of Mars indicated there were both clouds as well as surface features. And, secondly, that the planet appeared to have white spots at its poles indicating polar caps. It was to be some years before these conclusions were confirmed by more powerful telescopes than Maraldi possessed – though his drawings made in October 1704 indicate how close he was to the truth.

Confirmation that the white spots *were* accumulations of ice and snow was made by an English astronomer, William Herschel (1738–1822). His work in the field of astronomy was to lead to the discovery of the planet Uranus, two satellites of Saturn, and a host of new information about our solar system and the Milky Way, as well as a knighthood for himself.

Sir William's main observations of Mars were between 1777 and 1783, and these he reported in the *Philosophical Transactions* of 1784 under the grand title, 'On the remark-

19

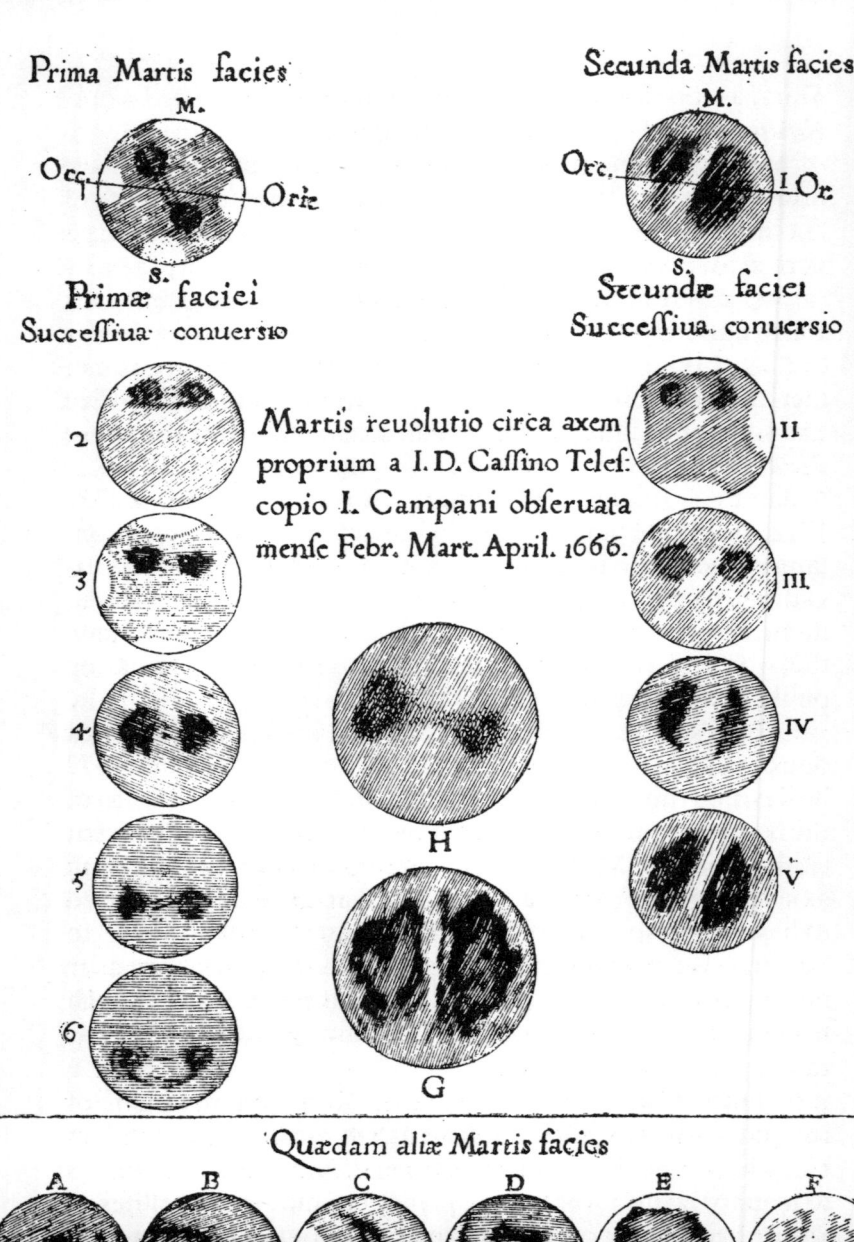

Prima Martis facies

Secunda Martis facies

Primæ faciei
Succeſſiua conuersio

Secundæ faciei
Succeſſiua conuersio

Martis reuolutio circa axem proprium a I. D. Caſſino Teleſcopio I. Campani obſeruata menſe Febr. Mart. April. 1666.

Quædam aliæ Martis facies

An array of Mars drawings made by Giovanni Cassini in 1666.

able appearances at the polar regions of the planet Mars, the inclination of its axis, the position of its poles, and its spheroidical figure; with a few hints relating to its real diameter and atmosphere.'

Like Christiaan Huygens, Sir William believed that Mars had an atmosphere and, consequently, inhabitants. 'These inhabitants,' he wrote, 'probably enjoy conditions analogous to ours in several respects.'

The English astronomer was the first man to note the colour changes which occur on Mars, and as a direct result of his detailed study he was able to establish when spring occurred on the planet. As more than one historian has noted, Sir William's report in the *Transactions* was one of the most important published about the red planet before the nineteenth century. (In fairness it has to be said that Herschel was not the first Englishman to write about Mars. This honour belongs to Robert Hooke (1653–1703), the chemist and physician, as well as a contemporary and rival of Sir Isaac Newton, who observed the planet in the year 1666 and reported his somewhat commonplace findings to the Royal Society the following year.)

Another important essay written shortly after this was, however, to remain unpublished for some time, thereby obscuring for a number of years the important contribution to Mars lore of an amateur German astronomer named Johann Hieronymus Schroter (1745–1816). Schroter was by occupation the chief magistrate at Lilienthal near Bremen, but such was his passion for astronomy that he obtained a 7-inch telescope built by Herschel and created an observatory where he devoted thousands of hours to the study of the planets and, in particular, Mars, between the years 1785–1802. The title of his essay, which was not published until 1881, speaks volumes for its contents: *Areographische Beitrage zur genaueren Kenntnis und Beurteilung des Planeten Mars* (Aerographic Contributions towards a Better Knowledge and Understanding of the Planet Mars). The article was illustrated with dozens of the author's sketches.

The reason for the delay in the publication of this work was that when the Napoleonic Wars burst upon Germany,

Schroter's home and observatory were destroyed. Amazingly, however, his astronomical papers survived and found their way to the University of Leyden. It was there, when the Mars fever began to take hold of Europe, that the director decided the essay on the planet should be published to make the public aware of the German amateur astronomer's contribution. He, as much as anyone of his time, can be seen through his sketches to have already begun to appreciate that Mars was not just another planet, but almost a 'second Earth'.

This appreciation was heightened in 1840 when two private astronomers in Berlin produced the very first 'map' of Mars. The men, Wilhelm Beer and J. H. von Madler, had been studying the planet from their observatory in the Tiergarten for some years, and aided by the sketches of their distinguished forebears produced their startling drawing. Indefinite it may have been, controversial it certainly was, but the public were fascinated and bought copies in huge quantities.

In 1858 another Italian astronomer, Father Angelo Secchi, made in Rome what are widely believed to have been the first colour drawings of Mars, complete with a fine shading of green for the planet's dark areas, and yellow for the lighter parts. The whole glowed a pale red. This innovative map was followed by some excellent drawings by an English astronomer, Sir Joseph Lockyer, in 1862, and a series by the German, Frederick Kaiser, which he consolidated into a single globe in 1864.

Three years later, in 1867, another English astronomer, Richard Anthony Proctor (1837–88) who had been busy determining the rotation of Mars, suddenly startled the world of astronomy by publishing a map of the planet giving all the main features *names*! For these he selected the names of other astronomers, both dead and alive, and although the quality of the map was undeniable, there was no little dissatisfaction among the fraternities of astronomers around the world to find that all the biggest features had been named after Englishmen! (See facsimile here.) And while there was no questioning the stature of men like Herschel, Cassini and Maraldi, what justification on Earth (or Mars, for that

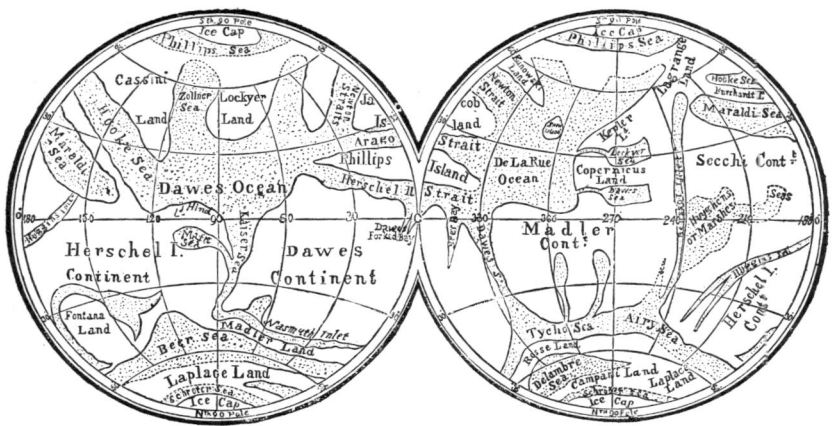

The controversial map of Mars by Richard Proctor, complete with its bizarre names (1867).

matter) was there in calling a huge landmass after a somewhat obscure English clergyman-cum-skywatcher, the Reverend William Dawes?

Perhaps it surprised no one that other maps appeared soon afterwards, offering their own alternative names for the large territories. One of the most striking of these originated from France in 1876 and was the handiwork of an astronomer and writer named Camille Flammarion who, as we shall see shortly, was to play a significant role in the next crucial development in Mars' history. Interestingly, Flammarion retained Proctor's basic idea of using the names of astronomers for the localities – he merely changed the emphasis and importance of certain of them.

More important still, though, was the map devised by an Italian astronomer named Giovanni Schiaparelli. It was important not just because it solved the nationalistic and linguistic problems by naming all the features in Latin, a factor acceptable to everyone then, and still in use today. No, it clearly depicted on the surface of the planet a criss-cross of lines, to all intents and purposes looking like rivers or even roads.

It was those lines which, in the words of the space historian, Willy Ley, 'suddenly made Mars the most popular of all planets, as it still is.' They seemed to hint at what men had only previously dreamed about. If the network was real, and the indications that it had been *created* were substantiated, then was the red planet inhabited – and by intelligent beings?

In fact, the appearance of Schiaparelli's map began what we can now see to have been a new epoch in Mars' history. An epoch which, as I shall next explain, created almost as many mysteries about the red planet as it resolved. . . .

TWO

The Legend of the *Canali*

THE YEAR 1877 marked the beginning of a new epoch in Mars observation. In the space of just twelve months two astronomers on opposite sides of the Atlantic changed Martian history irrevocably with the discovery, firstly, of some strange, artificial-looking lines running across the red surface, and secondly, with the revelation that the planet was being circled by two small moons. The mysteries which these announcements generated have remained familiar to this day, though the explanation of both has now been resolved.

During 1877 the opportunity for astronomers to observe Mars was extremely favourable, the planet being at its nearest approach to the Sun and the Earth. It was an opportunity seized upon eagerly by the director of the Brera Observatory in Milan, Giovanni Virgino Schiaparelli (1835–1910), who had been fascinated by the light and dark reddish areas of the planet for many years. What he observed through his 8.5-inch refractor telescope on a clear summer night was amazing in itself but sensational in its implications. For clearly defined across the huge, brilliant 'continent-like' areas was a perfect network of what he thought were very narrow dark lines, mostly straight, running in all directions and connecting the dark 'seas'.

'*Canali!*' the stunned Schiaparelli gasped under his breath. It was an observation that was to have profound and far-reaching effects.

In the report of his findings which the astronomer later published he was as specific as he could be about what he had observed. 'These lines,' he said, 'run from one to another of the dark spots on Mars, usually called seas, and form a very

25

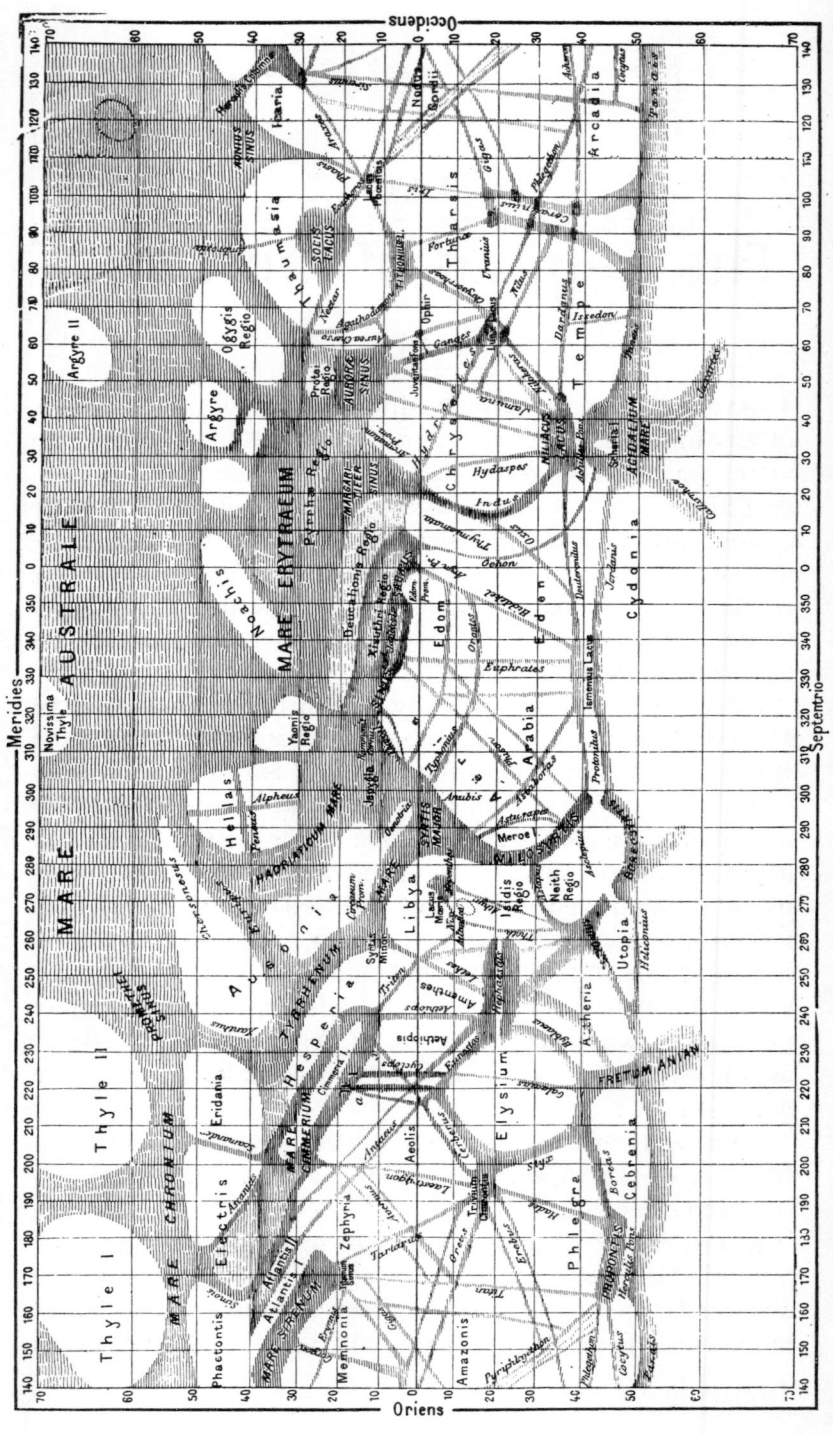

One of Schiaparelli's earliest maps showing the *canali* on Mars

well-marked network over the bright part of the surface. Their arrangement seems constant and permanent . . . and they do not at all resemble the winding courses of our streams. Some of the shorter ones do not attain 300 miles; others extend for thousands. Their number could not be estimated as less than 60.

'Some of the lines are easy to see, others are extremely difficult, and resemble the finest thread of a spider's web drawn across the disk. The colour is sometimes as dark as the seas of Mars, but often it is brighter. Each *canali* terminates at its two extremities either in a sea or in another *canali*.'

Along with this announcement, Schiaparelli – also an expert on planetary drawing – published a chart of the strange lines. It was greeted with a mixture of amazement, incredulity and even ridicule, among astronomers and laymen alike. And it was when the astronomer's word *canali* was translated into English incorrectly as 'canals' and not 'channels' of water as he had originally intended, that mankind's view of Mars was dramatically altered. For the implication of such a premise was that the lines were the work of life forms intelligent enough to have constructed massive waterways! For centuries, romantics had dreamt that beings *might* dwell on other planets in the solar system; now a respected scientist was apparently offering the proof they actually *did*!

With the passage of time, it has proved impossible to establish just how the mis-translation occurred, or why Schiaparelli did nothing to correct the misunderstanding. The evidence is that he must have come to accept the idea, for some years later he wrote again of the strange Martian lines, 'Their singular aspect has led some to see in them the work of intelligent beings. I am very careful not to combat this supposition, which contains nothing impossible.'

Others, as we shall see in a minute, also enthusiastically embraced this conviction.

The second major discovery of 1877 came hot on the heels of Schiaparelli's sighting, but from another observatory several thousands of miles away across the Atlantic in Washington. Here a dedicated astronomer named Asaph Hall (1829–1907), who had been quietly observing the heavens, suddenly announced that like the Earth and its moon, Mars also had a satellite. Indeed, not one but *two*!

Hall had also taken advantage of the proximity of Mars to try and confirm the long-held suspicion that the red planet had a moon of its own. He was well-equipped to search for the elusive body, the Washington Observatory then having the largest refracting telescope in the world at its disposal, a highly sophisticated 26-inch piece of equipment. With its help, Hall detected one faint object circling Mars on the night of August 11 and then – much to his surprise – as he continued his observations, a second on August 16.

There was precisely the same public excitement for Hall's discovery as had greeted Schiaparelli's. There could, though, be no disputing the actuality of the two satellites, and they were suitably named after the two attendants of the war god, Mars. The furthermost was called Deimos (Terror) and the inner, Phobos (Fear).

Just as there is a mystery around the *canali* of Mars – so there is one about the planet's two little moons. For there is evidence that they may well have been sighted much earlier, and not by an astronomer, but by Jonathan Swift (1667–1745) the great English satirist and writer of the classic novel, *Gulliver's Travels* (1726). In that book he presents what is an uncannily accurate description of Mars' two moons. I quote:

'The astronomers of Laputa have likewise discovered two lesser Stars or Satellites, which revolve about Mars, whereof the innermost is distant from the Centre of the primary planet exactly three of his Diameters, and the outermost five: the former revolves in the Space of ten hours, and the latter in twenty-one and a half; so that the Squares of their periodical Times are very near in the same Proportion with the Cubes of their Distance from the Centre of Mars, which evidently shows them to be governed by the same Law of Gravitation that influences the other heavenly Bodies.'

As there was certainly no telescope in existence in Swift's time able to see either of the Martian moons, this is surely an uncanny piece of prophecy. The fact that he should also have given such exact details of the positions of the moons makes it doubly so. No totally satisfying solution to this mystery has ever been advanced, although astronomers are generally agreed that it was just a piece of pure guesswork based on the then known facts about the universe in general and Mars in particular. A less likely explanation is the one I recall being advanced in 1964, that the great satirist knew all about the two moons because he was himself a Martian! According to this theory, he had arrived on Earth in a flying saucer on 7 November 1666!

If we discard that bit of nonsense and once again return to Schiaparelli's discovery of the *canali* in 1877, we find that the idea of there being inhabitants on Mars was not altogether new. Indeed, the French astronomer-writer, Camille Flammarion (1842–1925), eagerly seized on the Italian's chart to further substantiate claims he had been making for over a decade about the possibility of life in the solar system. He, too, was no wild-eyed romantic, but a practised student of the heavens who had been at work in the Paris Observatory since 1858, and was to crown his astronomical achievements by later founding the observatory at Juvisy in 1883. He tells us how he came to arrive at his conclusion that there might be life on Mars in his book, *Popular Astronomy*, published in 1894:

'The geographical knowledge which we now possess of the planet Mars is sufficiently advanced to enable us to draw a general map; this several astronomers have already done. I may mention that this neighbouring planet has always particularly interested me since the time when I wrote my first work on *Les Habitants de l'Autre Monde* (1862), because it was the first to bear witness to the truth of that great and sublime doctrine, by the light of which life and soul are scattered through the universe, instead of the solitudes in which float the material and immaterial masses of ancient astronomy.

'I drew, in 1876, a geographical planisphere of the planet, constructed from a comparison of maps and drawings previously made, and for which, in addition to my own observations, I used more than a thousand drawings made since the year 1636 – that is to say, since the first telescopic observations of the planet. Since 1876 the science has made further advances – so how long before we shall know perfectly the

Martian geography? When shall we distinguish the great cities of this neighbouring world? Sceptics may smile, as they smiled in the time of Copernicus and Robert Fulton (the American canal engineer and pioneer of steam navigation); but he who has confidence in progress does not despair of such a result.'

Flammarion devoted much of his energies in the remaining years of the nineteenth century to proving there *was* life on Mars, as the titles of his now rare and sadly neglected works bear eloquent witness. For instance *La Pluralité des Mondes Habités* (The Plurality of Inhabited Worlds, 1868), *Les Terres du Ciel* (The Worlds of the Sky, 1878) and most emphatic of all, *La Planète Mars et ses Conditions d'Habitabilité* (Mars and its Conditions of Habitability, 1892). He also wrote a highly imaginative novel of the life of the Martians entitled *Stella*, published in 1877, and took the work of Schiaparelli even further in his detailed map, *Carte de la Planète Mars*.

Important though Flammarion's popularizing of the idea of an inhabited Mars was, the man who really underlined the notion was the extraordinary American astronomer, Percival Lowell (1855–1916). Lowell, in fact, was a wealthy businessman who quit this career to follow his hobby of astronomy, and through his obsessional interest in Mars provoked a controversy seldom equalled in scientific circles.

Although, as I say, Lowell was enabled to pursue his interest in astronomy as a direct result of his wealth, he was also a brilliant mathematician, writer and lecturer who presented his case with such style and command of data, that his conclusions – though now largely disproved – still make for the most fascinating reading. That he attracted such a ready public following was also doubtless partly due to the fact that his early business travels in Japan had enabled him to produce several books which opened up the then little-known and mysterious East to Western readers. When his interest turned to Mars, he was able to perform the selfsame function for another mysterious world – this one in the heavens.

Because of the limited capabilities of astronomical telescopes then in existence in America, Lowell's first action was to build his own observatory with a massive new scope. For the site he picked the very dry and uniform climate of Flagstaff in Arizona, and there erected a 24-inch refractor. He took into his employ a small staff of experts in astronomy and launched an intensive campaign of observation of Mars, which was to continue throughout the remainder of his life. Today, the observatory he founded is still in operation, while the vast files of data he assembled and the books he wrote to substantiate his belief in the canals of Mars are also widely referred to. He is credited, too, with having done much of the research which lead to the eventual discovery of the planet Pluto.

Percival Lowell's first book about his fascination with the red planet was published in 1896 under the simple title, *Mars*. It was a record of the observations he and his team had made at Flagstaff between May 1894 and April 1895. Curiously – for the general public was still eager for more information about Mars – it had little impact. The book did, though, clearly indicate the direction in which Lowell was heading, as his four main conclusions show.

'*Firstly*,' he wrote, 'the broad physical conditions of the planet are not antagonistic to some form of life. *Secondly*, that there is an apparent dearth of water upon the planet's surface and, therefore, if beings of sufficient intelligence inhabit it, they would have to resort to irrigation to support life. *Thirdly*, that there turns out to be a network of markings covering the disc precisely counterparting what a system of irrigation would look like; *Fourthly*, and lastly, there is a set of spots placed where we should expect to find the lands thus artificially fertilized, and behaving as such constructed oases should.'

All this, Lowell was prepared to concede, 'may be a set of coincidences, signifying nothing.' But, he concluded, 'the probability points the other way. And as to details of explanation, any we may adopt will undoubtedly be found, on closer acquaintance, to vary from the actual Martian state of things; for any Martian life must differ markedly from our own.'

It was Lowell's second book, *Mars and its Canals*, published in 1906, which astounded everyone. In a nutshell, he declared that the lines on Mars were canals, built by intelligent beings across the smooth face of the planet in order to irrigate it. The red planet was also a dying world, he said, it had reached a stage in its history in which the desert was now dominant. Channelling the precious water from wherever it still existed – in the main from melting polar caps – was the planet's last chance of survival.

In a nice gesture to the man who had first fired his interest in Mars, Lowell dedicated the book to Schiaparelli, 'The Columbus of a New Planetary World' and he once more summarized the findings of his observations in the years since his earlier work.

'That Mars is inhabited by beings of some sort we may consider as certain as it is uncertain what those beings may be,' he wrote. 'The theory of the existence of intelligent life on Mars may be likened to the atomic theory in chemistry in that in both we are led to the belief in units we are alike unable to define.'

Lowell then turned his attention to the canals. 'Apart from the general fact of intelligence implied by the geometric character of their construction, is the evidence as to its degree afforded by the cosmopolitan extent of the action. Girdling their globe and stretching from pole to pole, the Martian canal system not only embraces their whole world, but is an organized entity. Each canal joins another, which in turn connects with a third, and so on over the entire surface of the planet. This continuity of construction posits a community of interest . . . and the thing that is forced on us in conclusion is the necessary intelligent and non-bellicose character of the community which could thus act as a unit throughout its globe.'

And with the final admonition of a prophet, he added, 'On our world we are able only to study our present and our past; in Mars we are able to glimpse, in some sort, our future.'

The culmination of all Lowell's work was published in the emphatically titled third book, *Mars As the Abode of Life* which appeared in December 1908, a short while before his death, and which re-emphasized all his earlier convictions with still more impressive observation. By this time, he was a man lauded with public honours, and a fellow of several of the world's leading astronomical societies. His books, too, were destined to become classics of modern science, a fact upon which the modern astronomer, Robert S. Richardson would write in his work, *Mars* (1965):

'They are required reading for everyone interested in planetary exploration. It doesn't matter in the least whether you agree with Lowell or not. If for no other reason they should be read for their style. Scientific papers today are written in a style that might be called Neo-Robot. They sound as if they were all written by the same person; or rather, as if they were all written by some machine. But out of Lowell's old-fashioned circumfluent prose there come passages of genuine emotional content.'

There was emotion in plenty, mixed with a very evident conviction, to be found in that last book, *Mars As the Abode of Life*. Take, for example, this further attempt to substantiate his earlier point about the comparisons which could be drawn between the histories of Earth and Mars.

'It is the planet's size that fits it thus for the role of seer,' wrote Lowell. 'Its smaller bulk has caused it to age quicker than our Earth, and in consequence it has long since passed through that stage of its planetary career which the Earth at present is experiencing, and has advanced to a further one, to which in time the Earth itself must come if it be not overwhelmed beforehand by other catastrophe. In detail, of course, no two planets of different initial mass repeat each other's evolutionary history; but in a general way they severally follow something of the same road. . . .

'This ageing in Mars' condition must have its effect upon any life it may previously have brought forth,' he continued. 'That life at the present moment would be likely to be of a high order. For whatever its actual age, any life now existent on the planet must be in the land stage of its development, on the whole a much higher one than the marine.

'The struggle for existence in their planet's decrepitude and decay would tend to evolve intelligence to cope with circumstances growing momentarily more and more adverse. But, furthermore, the solidarity that the conditions prescribed would conduce to a breadth of understanding sufficient to utilise it. Inter-communications over the whole globe is made not only possible, but obligatory. This would lead to the easier spreading over it of some dominant creature – especially were this being of an advanced order of intellect – able to rise above its bodily limitations to amelioration of the conditions through exercise of mind. What absence of seas would thus entail, absence of mountains would further. These two obstacles to distribution removed, life there would tend the quicker to reach a highly organised stage. Thus Martian conditions themselves make for intelligence.'

The fervour of Lowell's rhetoric compels him to bid farewell to his reader in these ringing tones: 'Thus, not only do the observations we have scanned lead us to the conclusion that Mars at this moment is inhabited, but they lead us to the further one that these denizens are of an order whose acquaintance is worth the making. Whether we ever shall come to converse with them in any more instant way is a question upon which science at present has no data to decide.

'But a sadder interest attaches to such existence: that it is, cosmically speaking, soon to pass away. To our eventual descendants life on Mars will no longer be something to scan and interpret. It will have lapsed beyond the hope of a study or recall. Thus to us it takes on an added glamour from the fact that it has not long to last. For the process that brought it to its present pass must go on to the bitter end, until the last spark of Martian life goes out. The drying up of the planet is certain to proceed until its surface can support no life at all. Slowly but surely time will snuff it out. When the last ember is thus extinguished, the planet will roll a dead world through space, its evolutionary career forever ended.'

As we shall see later in this book, there is rather more to these poignant words than might at first seem evident.

Before leaving Lowell's extraordinary work, there is one further observation he makes which I would like to quote. It is a singular vision of what man *might* experience when actually landing on Mars. For if our spacecraft have disproved the existence of the ancient canals he so fervently championed, they have, conversely, confirmed more than a few of the elements he here describes.

'Personal experience of Mars would take on a character akin to the grotesque,' he writes. 'Everything there would become unnaturally light: lead would weigh no more than stone with us, stone than water, each substance appearing to be transmuted into something other than itself. It would prove at once a world imponderable, etherealised. Our actions would grow grandific. For with little effort we should accomplish the apparently impossible, endowed with an effectiveness increase sevenfold. Lastly, everything would flow with hesitant and lazy current, and falling bodies sink with graceful moderation to the ground. After our first paranoiac wonder, it would certainly impress us as a world as slow as it was flat.'

As I have indicated, there were during Lowell's time plenty of sceptics about the idea of life on Mars, complete with its canals and super-intelligent beings. Why, they asked, had no one seen these strange marks before? And more particularly, why was it that many observers armed with telescopes far more powerful than Schiaparelli's and every bit as good – if not actually better – than Lowell's could see absolutely *nothing* on the planet's surface?

The distinguished American astronomer, Edward Emerson Barnard (1857–1923), who made a systematic photographic study of the heavens and had already discovered the fifth satellite of Jupiter in 1892, voiced the opinions of a good many people when he stated in 1894:

'I have been watching and drawing the surface of Mars. It is wonderfully full of detail. There is certainly no question about there being mountains and large, greatly elevated plateaus. But to save my soul, I can't believe in the canals as Schiaparelli and Lowell draw them. I see details where they have drawn none. I see details where some of their canals are but they are not straight lines at all. When best seen, these canals are very irregular and broken up – that is, some of the regions of their canals.

'I verily believe,' Barnard thundered, 'for all the verifications – that the canals as depicted by Schiaparelli are a fallacy and that they will be proved so before many oppositions are past.'

Schiaparelli's fellow countryman, E. M. Antoniadi, was just as sceptical when, aided by a much more powerful telescope than either of his colleagues, he wrote in 1909: 'At the first glance through the 32-inch telescope on September 20 this year, I thought I was dreaming and scanning Mars from its outer satellite. The planet revealed a prodigious and bewildering amount of sharp or diffused natural irregular detail all held steadily; and it was at once obvious that the geometrical network of canals discovered by Schiaparelli was a gross illusion. Such detail could not be drawn; hence, only its coarser markings were recorded in the notebook.'

It is interesting to note that both these two observations conform very neatly with what we know about the appearance of Mars. And they underline the opinion of more recent students of the story that the canals of Mars were almost certainly due to the eye's penchant for order. For it is much simpler to draw disconnected fine detail as a few lines, joining them up, than to put down all the irregular spots observed in an instant of good seeing.

As an astronomer with whom I discussed this phenomenon explained it to me, 'There is no question that the straightness of the lines is due to intelligence. The only question concerns *which* side of the telescope the intelligence is on!'

Interestingly, the man who effectively settled the issue of Mars and its canals – though, of course, the legend has persisted to this day – was actually an Englishman who was neither a romantic nor an astronomer. He was Alfred Russel Wallace (1823–1913), the naturalist whose travels in the Amazon basin and Malay Archipelago resulted in a memorandum to Charles Darwin which crucially aided the formulation of Darwin's seminal work, *The Origin of Species*. Wallace himself later published his own book, *Contributions to the Theory of Natural Selection* (1870). He was a vastly intelligent and rational man, and though well advanced in years when the Mars controversy broke, contributed a short work which has since been described as the 'high point of the intellectual discussion on Mars at that time'. The work began its life as a review of Lowell's *Mars and its Canals*, but such was the popular and scientific interest that greeted Wallace's conclusions that he expanded the argument into a book, *Is Mars Habitable?* Published in 1907, when he was 83 years old, it was an astonishingly far-ranging work which, had the grand old man lived longer, would probably have been the stepping-stone to still more prophetic observations on the mysterious red planet.

The core of Alfred Wallace's argument was a direct attack at one of Lowell's basic premises that Mars had a mild temperature ('equal to that of the South of England', he quoted the American astronomer as saying) which enabled the flow of water through the supposed canals. Wallace deducted – correctly as it has transpired – that the planet in fact had a temperature well below the freezing point of water. Therefore what was the point of having canals? And the *canali* – he further argued – were in reality just simple geological faults in the surface, and their apparent straightness was solely in the imaginations of Lowell and Schiaparelli.

On the last page of his book, the great naturalist summarized his case splendidly.

'All physicists,' he wrote, 'are agreed that, owing to the distance of Mars from the sun, it would have a mean temperature of about −35 °F even if it had an atmosphere as dense as ours. But the very low temperatures on the Earth under the equator, at a height where the barometer stands at about three times as high as on Mars, proves that from the scantiness of atmosphere alone Mars cannot possibly have a temperature as high as the freezing point of water; and this proof is supported by physicist Samuel Langley's determination of the low *maximum* temperature of the full moon. The combination of these two results must bring down the temperature of Mars to a degree wholly incompatible with the existence of animal life.'

To this he added, 'The quite independent proof that water-vapour cannot exist on Mars, therefore, means the first essential of organic life – water – is non-existent. And the conclusion from all of these independent proofs, which enforce each other in the multiple ratio of their respective weights, is therefore irresistible – that animal life, especially in its higher forms, cannot exist on the planet. Mars, therefore, is not only uninhabited by intelligent beings such as Mr Lowell postulates, but is absolutely UNINHABITABLE.'

Admiration for the deductions of Alfred Russel Wallace should not, however, be allowed to obscure totally the importance of Percival Lowell's work – for he did bring great dedication and enthusiasm to his observation of Mars, and undoubtedly his work did much to inspire the public debates which continued thereafter in learned journals, weekly magazines and, most particularly, in the minds of writers of what we now call 'science fiction'. The idea of the ancient and mysterious red planet being the home of intelligent beings was enough to fire the imagination of any storyteller. And the writings of one of these pioneers, H. G. Wells, was to have far-ranging influence in the beginnings of what has now become the race to reach Mars.

It is a fact, of course, that Mars had been featured in stories of imagination for a great many years before the dawn of the twentieth century. The German scholar Athanasius Kircher (1601–80) was one of the very first of these, to be followed by the Swedish theologian Emanuel Swedenborg (1688–1772) and the nineteenth-century British clergyman, W. S. Lach-Szyrma, who in 1874 published *Aleriel: or, A Voyage to Other Worlds*, which gave readers a vivid if highly romantic view of Mars. The success of this book led to another imaginative series, *Letters from the Planets*, published in *Cassel's Magazine* in 1887–93.

The first work to be wholly devoted to Mars was *Across the Zodiac* by the English poet, novelist and historian Percy Greg (1836–89). Written in 1880 and now regarded as an important if neglected early SF novel, the story concerns the flight to Mars of a huge spacecraft powered by an anti-gravity force called 'apergy'. The planet proves to be populated by a race of scientifically advanced people, and the hero is forced into polygamy, becomes involved in religious struggles, and accidentally kills some of his family with his imported germs, before finally escaping in his spacecraft. Though little read today, *Across the Zodiac* provided the model for many novels to follow.

Nine years later saw the appearance of another Martian adventure in *Mr Stranger's Sealed Packet* by an English mathematician, Hugh MacColl, which also described a journey to Mars in an electrically-controlled, anti-gravity spacecraft, though this time two races of inhabitants are found – both, curiously, of Earth origin! One of these races has attained a Utopian ideal which puts it into conflict with the other. Although the work was not the equal of Greg's novel, it proved highly popular with the reading public, and science fiction expert John Eggeling has suggested that it might have been an influence on H. G. Wells, 'especially in its account of the death of a Martian female through exposure to bacteria in the Earth's atmosphere.'

The next year, 1890, a third British writer, Robert Cromie (1857–1907), who lived in Belfast and was an avowed admirer of Jules Verne, again utilized the idea of a space mission to Mars which discovers a group of humans living in Utopian conditions in his work, *A Plunge into Space*. The 'Steel Globe' in which the intrepid voyagers travel is also interesting in being something of a forerunner of the modern communications satellite in appearance and functions.

Despite all the work which Lowell had put into the promotion of the 'life on Mars' idea, it was not until 1890 that the first major American novel on the theme was written by a Philadelphia clergyman, Mortimer D. Leggett (1850–1915). Perhaps not surprisingly, *A Dream of a Modest Prophet* had a strong Christian slant to it, but was interesting in its portrayal of a Martian civilization which has undergone similar social and spiritual developments to those on Earth, but is now some 3,000 years in advance of us.

The following year saw the appearance of one of the most successful early American Mars books, *The Man from Mars* by one Thomas Blot, which sold at least three editions. The author was in actual fact the San Francisco astronomer, William Simpson (1861–1928), who contributed a number of articles on Mars to learned journals before writing this tale of a Martian who visits a reclusive astronomer and tells him all about the advanced civilization on his native planet. Despite the book's success, it was nowhere near as accomplished a work as Robert D. Braine's *Message from Mars*, published in New York in 1892. This unusual tale is noteworthy as being the first to describe the Martians as non-human in form, and in its idea of attempting to communicate with the red planet by way of radio it presaged a number of such attempts which were actually made in the twentieth century. Although we shall look at these endeavours later, unfortunately we know nothing about the ingenious Robert Braine.

Another influential novel, *Journey to Mars* by Gustavus W. Pope (1858–1930), appeared in 1894. Pope, a New York doctor, subtitled his book, 'The Wonderful World; Its Beauty and Splendour; Its Mighty Races and Kingdoms; Its Final Doom' and described a flight to Mars in an electrically operated spaceship which uses fields of magnetic force as its power. But once on the planet, the story becomes a vivid mixture of romance and science about a decayed, once advanced civilization in which the inhabitants war on each other with a mixture of old-fashioned weaponry – such as swords and staves – and the most advanced and destructive scientific hardware! The central character is an American officer who falls in love with a Martian princess, and more than one science fiction expert believes this work to have been a considerable influence on Edgar Rice Burroughs and his famous series of Martian stories.

Ahead of its time, too, was Professor Willis Mitchell's *The Inhabitants of Mars* (1895). Mitchell, a Massachusetts scientist and apparently a great admirer of Lowell, envisioned a society which had already mastered its weather, lived in sophisticated underground settlements, journeyed in electrically driven vehicles and used birth control as well as prenatal sex selection to regulate its society. It was a book that raised hackles in various religious circles and was vigorously attacked in a number of newspapers. Copies today are of the utmost rarity.

The fever of Mars fiction was also evident in Europe, where *From India to the Planet Mars* by Theodore Flourney, a professor of psychology at the University of Geneva, became a best-seller in France in 1894 and was later translated into English for publication on both sides of the Atlantic. The book claimed to be a revelation about life on Mars obtained through a number of seances. It is perhaps more highly regarded in the field of psychic research than for having any real relevance to Mars!

More important than this book – and indeed still highly regarded as a major early work of science fiction – is *Auf Zwei Planeten* (Two Planets) by the German philosopher and science historian, Kurd Lasswitz (1848–1910). Apart from being the first novel to feature the importance of interplanetary relations, it marked Lasswitz as the German equivalent to Jules Verne in France and H. G. Wells in England. The story described the effect on mankind when a Martian satellite is discovered over the North Pole. Attempts to drive off this alien beachhead proved futile against the superior Martian culture and weaponry, and the Earth is put under a benign protectorate. Then, strangely, as the Earth people benefit from this influence, the Martians become decadent, and an ultimate settlement between the two worlds is destined to give the Earth a Utopian future.

Lasswitz' extraordinary novel clearly took some of its ideas about life on Mars from Lowell's books, while the technology was of such an imaginative kind as to earn the author his comparison with Verne and Wells. Not surprisingly, *Two Planets* proved highly successful and was deeply influential on at least two generations of German youth. It has rarely been out of print since its initial publication, and the most recent translation (1971) was made by none other than the eminent rocket expert, Wernher von Braun. During his lifetime, Lasswitz explained that he had gone to great pains to make the story as authentic as possible.

'Many inferences about the future can be drawn from the historical course of civilization and the present state of science,' he said, 'and analogy offers itself to fantasy as an ally.'

All of these novels I have mentioned were, of course, important in their own way. But one single work – H. G. Wells' Martian extravaganza, *War of the Worlds* which was first published in serial form in the British magazine, *Pearson's*, in 1897 – surpassed them all both in terms of its world-

41

wide popularity and its influence on our history.

The facts about the life of Herbert George Wells (1866–1946), the son of a struggling shopkeeper, who became a teacher and in the textbooks of science found the inspiration which was to make him one of the true greats of science fiction, have been so often repeated as to need no retelling here. The background to his fascination with Mars is, however, less well known.

It was in 1883 while he was at Midhurst Grammar School that he came across the beautifully illustrated translations of the works of Camille Flammarion and fell in love with the intriguing maps and diagrams of the mysterious red planet. He also found Flammarion's novel of life on Mars, *Stella*, quite breath-taking. Thereafter there developed in him an interest in Mars which remained to the end of his days.

The first literary result of this fascination was a short story called 'The Crystal Egg' which he wrote for *The New Review* early in 1897. This described life forms and scenes on the planet which are transmitted to Earth by means of a crystal egg sent from Mars. It is an intriguing little tale, but gives no real clue to the master work which was already hatching in the author's imagination.

The War of the Worlds, which appeared a few months later, changed the face of science fiction – as historian Brian Stableford has remarked. 'It cast a long shadow over the SF of the twentieth century,' he says. 'Wells' Martians, attempting to escape their due fate on a waterless, dying world, come as predatory Darwinian competitors to claim Earth from the human race. This novel firmly implanted in the popular imagination the image of the Martian as monster, and brought a new sensationalism into interplanetary fiction.'

The story also profoundly affected readers all over the world. And not just the English-speaking world. For *War of the Worlds* was translated into dozens of languages and even penetrated into the farthest reaches of pre-Revolution Russia. Here it was to affect radically the thinking of another scientifically-minded teacher named Konstantin Tsiolkovsky.

Far away in America, too, another younger man named Robert Goddard similarly pored, enthralled, over its pages and began to dream of finding a way of actually going to Mars.

Though these two men had no way of knowing it then, they were to prove the catalyst which would begin their respective countries' race towards the mysterious red world across the enormity of space. . . .

THREE

The Quest Begins. . . .

IT WAS ONLY a matter of months after the publication of H. G. Wells' sensational novel, *War of the Worlds*, that a translation – unauthorized as it happened – was available in Russia, and was eagerly read by a schoolteacher in the large provincial town of Kaluga. Konstantin Eduardovich Tsiolkovsky was a 40-year-old mathematics and physics master whose abiding interest outside his school work was space travel. For years – from his early twenties, in fact – he had been working on solving the problems of interplanetary travel, and apart from notebooks full of calculations and drawings of strange machines, he had also written a few fantasy stories about space flight. Like the Wells stories, too, there were elements of brilliant prophecy in what he had written, and he was understandably attracted to the work of the Englishman who, like him, was also a teacher. Prior to reading *War of the Worlds*, Tsiolkovsky had directed most of his attention to the idea of reaching the moon; this story fired his imagination with thoughts of the more distant yet infinitely more intriguing red planet.

Because the work of Konstantin Tsiolkovsky (1857–1935) was published mainly in obscure journals, and his country was preoccupied with the social turmoil which was to climax in 1917 in the Russian Revolution, it has only been in recent years that the true importance of this pioneer and the influence he has had on Soviet space plans have become appreciated in the West. Yet once we begin to study the available material, it is not hard to see why in Russia he is hailed as 'The Father of Cosmonautics'. He was certainly the first man to appreciate fully the significance of rockets and develop an

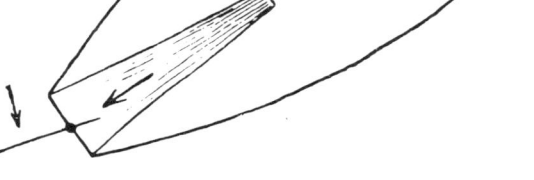

11. Поворачивание ракеты взрыва-
нием при наклонении руля. Вращение.

One of Tsiolkovsky's sketches for a jet-propelled
spaceship – reproduced from his notebook.

understanding of their use in space travel. He was also the
first to evolve theoretically the multi-stage principle in
rockets and the use of solar energy for some of the batteries in
the artificial satellites of the future. He was also – the
Russians claim – the first man to diagram an Earth satellite
and a space station. He is further credited with having outlined the basic method by which it is generally accepted the
Russians intend to reach Mars.

Tsiolkovsky was born in the village of Izhevskoye near
Moscow in 1857 of mixed Russian-Polish-Tartar blood. He
was a bright and lively child by all accounts until, tragically, at
the age of nine he contracted scarlet fever and became deaf.
This shattering blow caused the youngster to withdraw very
much into himself, and instead of seeking out playfellows he
spent his time reading the books in the voluminous library
assembled by his amateur inventor father. Here his love of
literature was born, and in particular his obsession with works
of science and the imagination. On the library shelves he
found serious scientific works on the history of rocketry as
well as the fantasies of the great Frenchman, Jules Verne.
Together these books shaped his destiny.

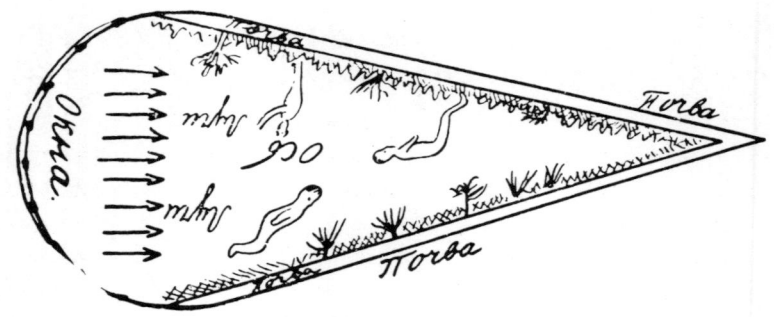

Another of Tsiolkovsky's prophetic drawings proposing a space station.

'There was nothing that engrossed me more,' he was to write years later, 'than the problem of overcoming terrestrial gravity and of making flights into space. The things I have brought about, the ideas that have surged through my brain! This was no longer fantasy, but exact knowledge based on the laws of nature; new discoveries and new compositions were in the making. But I was also attracted to science fiction. Many times I essayed the task of writing about space travel but wound up becoming involved in exact compilations and switching to serious work. Science fiction stories on inter-planetary travel carry new ideas to the masses. All who are occupied with this are doing good work; they excite interest, promote the working of the brain and bring into being people who sympathize with, and will in the future engage in, work on grand projects.'

Tsiolkovsky had begun to dream of 'grand projects' in his early teens when his deafness had made him unable to hear the teachers at school, and so he stayed at home teaching himself first mathematics and then physics. Experiments which he carried out with hydrogen balloons and flying machines with flapping wings, all of which he built in his father's workshop, indicated the direction his inquisitive mind was taking. Of perhaps the most ambitious project at this time – a large navigable balloon with a thin metal shell – he wrote:

'I was fascinated by the astrolabe which enables us to

measure distances to objects beyond our reach. I constructed a height-finder. Then, with the help of the astrolabe, I calculated the distance between our house and the fire-tower without going outdoors. The calculation showed it to be 1,200 feet. I then measured the distance with the measuring-rod and in this way verified my calculations. This made me believe in theory.'

It was by now obvious to Konstantin's father that his son deserved some kind of proper education, and consequently he scrimped together enough money to send him to Moscow with a modest allowance. There the boy's voracious appetite for facts caused him to spend rather more of his money on books than food. He invested in retorts for experiments, too, as well as quicksilver and sulphuric acid. Rocket power had taken control of his mind.

It seems curious in the light of what has happened in the twentieth century to realize that it took mankind such a long time to appreciate the importance of rockets. They had, after all, been in use for centuries for military purposes, the Chinese being perhaps the earliest to employ them, in the first century A.D. Roger Bacon had, of course, invented the improved form of gunpowder used for firing incendiary projectiles in 1242, with Sir William Congreve perfecting the idea of a destructive rocket manoeuvred by fins almost six hundred years later. Yet not until the nineteenth century did scientists begin to understand the true significance of rocketry.

Tsiolkovsky was one of the very first to do so. And soon the makeshift laboratory in his room began to resound to 'lightning flashes, peals of thunder, wheels revolving and all manner of strange sounds and lights' according to one account. To broaden his knowledge still further, he even took to attending scientific lectures – having devised for himself a tin ear-trumpet which served as a crude hearing aid!

'I was 17 when I first dreamed of the possibility of travelling beyond our own planet,' he was to explain years later. 'In an exercise book I drew home-made contraptions and I experimented with mice, chickens and insects to ascertain the effect that acceleration of gravity has on living organisms.'

But this study and experimenting was all very well – it did

not amount to a career. And so after consultation with his father, the young man decided on taking a course to become a schoolteacher. This he passed with flying colours in 1879, and as a result was appointed to teach maths, geometry and physics at the Borovsk District School in Kaluga, 45 miles south-west of Moscow. He proved an excellent, if occasionally absent-minded instructor, but undoubtedly found it increasingly difficult to put to one side the ideas of space travel which intruded on his thoughts whenever he was called upon to take a class.

Nothing could shake his obsession, and in 1883 he wrote his first major essay, 'Free Space', a development of the theories he had begun scribbling in his teenage exercise book. In this he analysed the various problems of mechanics which a man would encounter if he could penetrate interplanetary space, viz gravity and weightlessness. He also included details and a drawing for a rocket spaceship which he powered by what can only be defined as jet propulsion. It was the first of its kind in the world.

'Consider a barrel filled with highly compressed gas,' he wrote. 'If we open one of the taps the gas will escape through it in a continuous flow, the elasticity of the gas pushing its particles into space will also continuously push the barrel itself. The result will be a continuous change in the motion of the barrel. Given a sufficient number of taps (say six) we would be able to control the gas exhaust however we choose, and the barrel (or sphere) would describe a curved trajectory whose radius would depend on the barrel's velocity.'

The jet-propelled spacecraft also contained gyroscopes which could be revolved to change its direction in space.

The article was an astonishing achievement for a 25-year-old, for in it Tsiolkovsky revealed a basic understanding of reaction flight, and it has been claimed, with some justification I think, that had this discovery been applied it could well have led scientists towards space flight many years ahead of schedule. As it was, the article actually remained unpublished for a number of years, so revolutionary was it considered.

Perhaps because he sensed his ideas would *not* be acceptable as serious propositions, Tsiolkovsky began instead to write stories of science fiction into which he cleverly introduced the fruits of his thoughts and experiments. In 1887, for instance, he wrote a novelette entitled 'On the Moon' about a traveller's impressions of this barren satellite, and then in 1895 published 'Dreams of Earth and Sky' in which he proposed a ball-shaped rocket ship which is fired into space from a huge cannon. More remarkable than either of these tales, however, was 'Outside the Earth' a discussion between a group of scientists about the possibilities of interplanetary flight. In this Tsiolkovsky included himself as a man described enigmatically as 'The Russian'. In the course of the narrative, this man suffers the ridicule of his fellows as he outlines an ambitious and far-sighted project to conquer space. In the end, though, they are won over by the sheer brilliance of his argument. Almost a hundred years later, the modern reader likewise finds himself just as impressed. Take, for example, this description of the Russian's space rocket.

'The vehicle was a 20-metre-long cigar-shaped metal structure 4 metres in cross-section and standing on end. Numerous portholes let in a sufficient amount of light to the interior quarters. Three pipes ran down along its walls, protruding at the lower end. There were many mechanisms partly shielded by metal casings and large tanks of strange liquids. When mixed they produced a continuous, uniform blast the products of which escaped with tremendous force through the pipe nozzles in the lower part of the projectile. A row of knobs and various dials were intended for steering the projectile and varying the thrust.'

The Russian then explains what will happen after the fuel of 'oxy-hydrogen gas' is ignited and the spacecraft lifts off.

'We see that in a few seconds the projectile will reach extremely rarefied atmosphere,' he says, 'and several seconds later will be travelling in total vacuum. Assuming the average thrust force of the gases to be ten times greater than the total weight of the rocket and its pay load, we find that, given the most powerful propellants, it will consume its entire fuel supply in 160 seconds. By then, it will have risen to an altitude of 1,152 kilometres and attained a maximum velocity of

14,400 m/sec. This speed is sufficient to carry it away from the Earth, and all the easier will it be to reach any planet of our system.'

Next the persuasive Russian discusses the need to provide oxygen for the crew and – again with remarkable foresight – the necessity for an 'automatic pilot' in the craft, 'since in the unusual flight conditions a person might easily lose his presence of mind and fail to operate the necessary controls.' Our narrator explains:

'By common consent it was decided to feed the following instructions into the automatic pilot: the rocket would take off parallel to the plane of the equator at an angle of 25 degrees to the horizon in the direction of the Earth's rotation. During the first 10 seconds its velocity would increase rapidly to 500 metres. In passing through the atmosphere the rate of the acceleration would be low until the air was sufficiently rarefied. With the Earth's atmosphere left behind the velocity would again increase rapidly, the direction of the flight gradually changing until, at an altitude of 1,000 kilometres, the rocket would enter into circular orbit. The speed at that point should be such as to keep the vehicle in a circular path about the globe without approaching it. The automatic pilot, of course, could be switched off or re-set.'

Today this reads uncannily like the description of a typical space flight – but was actually written by a reclusive schoolmaster in the final years of the last century! And nor do Tsiolkovsky's powers of foresight end there. His 'Russian' also has observations to make on the condition of weightlessness which the crew of the spacecraft will experience – and has some things to say about the obvious necessity of space suits.

'When the combustion of the fuel is stopped and the rocket is no longer accelerated by the pressure of the exhaust gases,' he says, 'the relative gravity will disappear completely, regardless of the magnitude of the all-penetrating gravitational forces. The crew will then float in their atmosphere, neither falling nor pressing against the floor or obstacles. They'll be like fish in water with the advantage that they won't have to overcome the great resistance of water when they move.'

The narrator also correctly foresees the problems men will

have eating in space.

'Plates and decanters and the food itself would have to be fastened down,' he says. 'If you laid your spoon or fork on the table it would sail away towards your neighbour and he would have to be on the alert to prevent a fork poking his eye out or a knife from grazing his nose! When cut loose, crumbly food would scatter about and get into your nose, mouth, eyes, ears, hair and pockets. If you wanted a glass of water, the water would not pour. You could throw back your head to toss off a glass of wine, but the wine would fly out, break into several large globules and float away, wetting the clothes of diners or entering the mouth of someone who had no intention of drinking. . . .'

But the problem has been solved. 'The food will be kept in closed containers,' the Russian declares. 'Semi-liquid and liquid products will be pushed out of their containers by pumping in air and thence directly into the mouths of the diners.'

In the final section of this remarkable story, Tsiolkovsky describes a space suit, once again with uncanny accuracy.

'Some day we may have to land on planets with atmospheres unsuitable for breathing, either because of their composition or their extreme rarefaction. The same type of outfit is suitable for survival in a void and in rarefied or hostile atmospheres. A typical space suit will cover the whole body, including the head, and is gas and vapour tight, flexible, light and allows the body freedom of movement. It is strong enough to withstand the internal pressure of gases surrounding the body and the helmet has special flat visors. It has a thin, warm lining through which gas and vapour can pass. It has reservoirs for urine and other purposes and is connected with a special cylinder which provides an adequate supply of oxygen. Carbon dioxide, water vapours and other products excreted by the body are absorbed in special vessels. Automatic pumps continuously circulate the gases and vapours inside the suit through the permeable lining. Each man needs no more than a kilogram of oxygen a day. The suit has an eight-hour supply, and its total weight is 10 kilograms.'

On the evidence of these remarks, it is not hard to see why 'Outside the Earth' should be considered such an impressive piece of writing. Yet it was couched in fictional terms and hardly likely to catch the interest of the leading Russian scientists. The next phase of Tsiolkovsky's work was to present such an impressive case for his theories of jet propulsion in interplanetary travel that *no one* could deny his argument. This he did in January 1903 when he submitted to the *Nauchnoye Obozrenie* (Scientific Review) in St Petersburg an article entitled 'Exploration of Cosmic Expanse Via Reactive Equipment'. It was a long, complex and at times obscure piece in which he perhaps unwisely overelaborated the mathematical formulae in order to justify his beliefs about space rockets. Amidst all this verbiage, though, was perhaps his most interesting point of all – the view that mankind would ultimately find Mars an ideal planet for colonization. It was evidently a thought that had been generated from his enthusiasm for H. G. Wells' *War of the Worlds* – though Tsiolkovsky was convinced that the red planet did not harbour any creatures such as the English writer had described!

However, though the article was eagerly accepted by the editor of the magazine, its publication was fraught with difficulties. Firstly, it was so long as to have to be divided into two parts. Secondly, the editor apparently had last-minute reservations and added an apologetic comment that he felt 'the author's fantasy perhaps flies too far off.' And, thirdly, no sooner had the first instalment appeared in the May 1903 issue, than the magazine – for some time under suspicion by the Tsarist government for having political overtones – was closed down by the police and all copies on the premises, as well as all manuscripts, were seized. It is doubtful if more than a few hundred readers ever received their copy of that issue and certainly the journal never appeared again. (It was not, in fact, until 1911 that Tsiolkovsky was able to publish the second part of his essay in a flying magazine, *Vestnik Vozdukhoplavaniya* – Herald of Aeronautics – of all places!)

But, as we shall see, nothing should be allowed to obscure the importance of this essay – in particular its references to Mars. For during the course of it he proposes the use of future space stations as the jumping-off stage for the exploration of Mars and other planets. Let Soviet Academician S. Korolyev summarize Tsiolkovsky's argument.

'According to Tsiolkovsky,' he wrote in a tribute to mark the one hundredth anniversary of the experimenter's birth, 'such space stations should be made up from many rockets which, having reached the necessary speed and having expended their store of fuel, could be employed as parts of a future station. On such interplanetary stations as he suggested, it would be possible to grow plants for man's food. Man thus could stay on such stations for a long time. Possessing permanent space stations, he could facilitate the flight of cosmic rockets to distant destinations, also their landing on their return way. Such a station would give men an opportunity to store fuel for the rockets and all the necessities for humans. Connection with the Earth could be maintained by special rockets.'

If we now look at Tsiolkovsky's actual article, there are two specific quotes which make particularly fascinating reading.

'As these space stations increase,' he says, 'they will expand their industries and be able to build things without help from Earth. In time, the Earth will manufacture only propellants and rockets for carrying people. The rockets, after serving their purpose, will be able to return on fuel manufactured "up there".'

And later he claims: 'The use of these reaction devices will allow man to step onto the soil of asteroids; to form rings – with life upon them – around the Earth and Moon; to observe Mars at close quarters; to land on the satellites of Mars or even on Mars itself.'

'Mars' he concludes, 'presents a great challenge to us. But it will not run away. We will reach it in the course of some later expedition.'

Highly prophetic words in the light of what is happening today!

Despite the importance of his vision, Tsiolkovsky was destined to spend the rest of his days as a schoolteacher, both in obscurity and virtual penury. Although his ideas were examined from time to time by government and scientific officials, the establishment was more interested in the use of rockets for *military* purposes than interplanetary travel. The fact that his work only appeared in insignificant journals completely unknown outside Russia meant that the West was unaware of his existence for many years. As Professor Albert Parry, the American space historian, noted of Tsiolkovsky in 1960, 'Recognition came slowly and in pitiful doses, yet it did come. Before he died he saw his own Russian pupils and followers send up the first rocket of value to science.'

Professor Parry has said that although Western experts are at last prepared to give the Russian more credit than previously, 'they still doubt that in the more crucial areas of research and speculations, he was really ahead of America's Robert H. Goddard.'

And in that sentence lies the crux of the matter for – as we shall see – Goddard, like Tsiolkovsky, may also have been something of a dreamer, but he was also an intensely *practical* man. Not content with mere speculation and simple experimentation, he put his ideas for rocket flight into *practice* and achieved the satisfaction of seeing them actually take flight.

But we cannot deny Tsiolkovsky the honour of being the first to seriously propose space stations as in-between 'platforms' en route to Mars and other celestial bodies, and the first to discourse in some detail on the possible conditions under which men could live in space. In the Soviet Union he is considered the 'father' of the Sputnik and all subsequent Russian space achievements, and if in the West we may dispute some of the other grandiose claims attached to his name, those things we should allow.

Tsiolkovsky died in 1935 of cancer, bequeathing his notebooks, sketches and his little store of equipment to the Communist Party. In his honour, his home in Kaluga has now become a museum, and a bronze statue of him with a silver· model of a rocket as a backdrop stands nearby. His writings have also been collected from obscurity and are now freely available in Russia, though still hard to obtain in the West.

There are two footnotes to be added to Tsiolkovsky's life – one sad, the other happy. In 1920, H. G. Wells visited Russia and made a point of seeking out as many scientists as he could. The reclusive schoolmaster in Kaluga had, of course, no way of knowing that the man who had inspired a new dimension to his thinking was in the country. And, sadly therefore, they never met. One can only speculate on what they *might* have learned from each other had the opportunity arisen.

The happy occasion occurred in 1933 not long before Tsiolkovsky's death. That year the very first Russian-built rocket was sent up from a Moscow aerodrome. It reached a height of just a few miles, and despite his illness, Tsiolkovsky was beside himself with delight when the news was relayed to him.

However, the young friend who brought him the news was amazed that the old man who had dreamed of travelling the far reaches of space could get so excited about such an insignificant achievement.

Tsiolkovsky shook his mane of white hair. 'Remember the first flight of the aeroplane!' he reproached the boy. 'Not many understood then that a new era was dawning. The same is true of this first rocket-launching. It opens a new page in man's conquest of endless expanse – with the aid of rockets. The hour is not far off when Soviet rocket ships will rush into great airless expanses. And this will happen in our twentieth century.'

On his deathbed, the remarkable man was to add one final observation. 'For forty years I have been working on the reactive engine,' he said, 'and I thought that a journey to Mars would begin in a hundred years or so. But time perspectives change. I now believe that many of *you* will witness the first trans-atmospheric journey!'

Without quite realizing it, Konstantin Eduardovich Tsiolkovsky had spoken the words of inspiration which have driven the Russian people to the very threshold of Mars upon which they now stand. . . .

☐

Though they never met and lived on opposite sides of the world, there were several features in common about Konstantin Tsiolkovsky and his American counterpart Robert Hutchings Goddard (1882–1945). Both suffered early illness and retreated into the world of science. Both spent years in obscurity working on their theories which the passage of time has shown to be years ahead of their contemporaries. And, of course, both had been inspired in their work by *War of the Worlds*.

Goddard was born in the delightful rural Massachusetts town of Worcester in October 1882, just a year before Tsiolkovsky published his first major essay about reaction flight. But it was in 1903 that crucial events occurred in the lives of both men, as Frank H. Winter has described in his book, *Prelude to the Space Age* (1983):

'In Russia, the same epochal year that the Wright brothers flew above the dunes of Kitty Hawk, the deaf provincial schoolteacher Konstantin Tsiolkovsky published his classic article "Exploration of Cosmic Expanse Via Reactive Equipment". Earlier, in Massachusetts, the schoolboy Robert H. Goddard had dreamed of a trip to Mars after reading Wells' *War of the Worlds*. By the time of the Wrights' flight, young Goddard was determined to make a career of the study of "reactive motion".'

Despite the healthy environment into which he was born, Robert suffered one childhood illness after another, and so he poured all his energies into the study of science. He proved particularly adept at maths and physics and was soon employing this talent in searching for ways to make his dream of space travel a reality. Interestingly, while this research increased his admiration for the speculative fiction of H. G. Wells, it made him somewhat sceptical of the other great science fiction writer, Jules Verne. So much so, in fact, that he even went so far as to re-write completely *From the Earth to the Moon* with Verne's canon launcher turned into an eminently more feasible rocket ship!

It is evident from his notebooks that Goddard was only in his teens when he first began to speculate seriously about space travel. His first experiments were with balloons made of aluminium and filled with hydrogen, but when they proved too heavy to reach high altitudes he switched his attention to rockets. In an entry in his diary for October 1899 (when he was just 17), he commented about conditions in the upper atmosphere and interplanetary space, and conjectured that rockets would best provide the means of carrying instruments to study these regions.

Interestingly, a factor that had influenced Goddard's speculations had occurred when he had been pruning a cherry tree in his father's garden. Just as Newton had found his inspiration under an apple tree, so Goddard discovered his on a ladder. 'I was a different boy when I descended that ladder,' he was to note later. 'Life now had a purpose for me.'

Though Goddard proved one of the most able pupils at the Worcester Polytechnic Institute, it was in his spare-time writing that he most vividly revealed the ruling passion of his life. For by 1902 he had written an article entitled 'The Navigation of Space' in which he proposed the rocket as the most effective means of achieving this objective.

Unlike his Russian counterpart, though, Goddard was not content merely to speculate – he determined to put his theories into practice. And an amusing anecdote from the year 1908 – the year he graduated from the Worcester Polytechnic – records that classes were suddenly disrupted one morning by the sound of an explosion and the eruption of smoke from the basement of the Institute. It was caused by Goddard's very first static test of a small rocket!

Goddard had also been busy with his pen, too. He had prepared for publication another paper in which he suggested that the heat from radioactive materials could be used to expel substances at high velocities through a rocket motor, thus furnishing enough power for space flight. He submitted the draft to America's three leading scientific journals, *Popular Science Monthly*, *Popular Astronomy* and *Scientific American*, but all three turned it down flat. Quite simply, Goddard was forty years ahead of his time!

Nonetheless, the young man was not to be deterred in his beliefs. After his graduation (and despite the explosion!) Goddard remained at the Worcester Polytechnic as an instructor in physics while taking a post-graduate course at Clark University, ultimately obtaining his Ph.D. in 1911.

It was in 1909, while studying for his professorship, that the 27-year-old Goddard came to the conclusion that it would require the use of high energy propellants such as liquid oxygen and liquid hydrogen to achieve interplanetary flight, coupled with a rocket constructed on the multiple, or step, principle.

Then followed several years of intense theoretical research

and calculation. By 1913 he had evolved a theory based on the possible use of a smokeless powder for the rocket propellant as well as oxygen and hydrogen. The very next year Goddard began to put his ideas to the test.

To begin with, he conducted a series of tests with a standard 'black powder' rocket, using a ballistic pendulum to measure the impulse. From these experiments he learned that such rockets had an exhaust velocity of about 1,000 feet per second, and were operating with an efficiency as low as two per cent. Then he continued using heavy steel chambers of half-inch and one-inch diameter, fitted with expansion nozzles. With charges of smokeless powder fired by electric wire, he achieved exhaust velocities as high as 7,964 feet per second and operating efficiencies of up to 64 per cent.

His most important discovery, however, was that a rocket motor not only functioned in a vacuum, but also did so with greater efficiency than in an atmosphere. This he established by firing smokeless powder in a small steel chamber which was placed in a large evacuated steel pipe, the impulse being recorded by means of a recoil spring and a steel needle which scratched a line on a piece of smoked glass. At a stroke, he had established beyond question that a rocket ship *would* be able to operate in space.

On the verge of still greater discoveries, Goddard's work nearly foundered when the funds available for this naturally expensive work ran out. He urgently sought financial aid by submitting a report which he entitled 'A Method of Reaching Extreme Altitudes' to various technical bodies. Fortunately, the prestigious Smithsonian Institute in Washington happened to be interested at the time in high altitude devices for meteorological research, and though Goddard's work was clearly aimed at more ambitious targets, decided to advance him a grant of $11,000.

Goddard's paper, which was later published in 1919 by the Smithsonian Institute, is remarkable by any standards. In the first part, he outlines the theoretical requirements of rockets designed to reach various altitudes.

'The rocket,' he wrote, 'is ideally suited for reaching high altitudes, in that it carries apparatus without jar, and does not

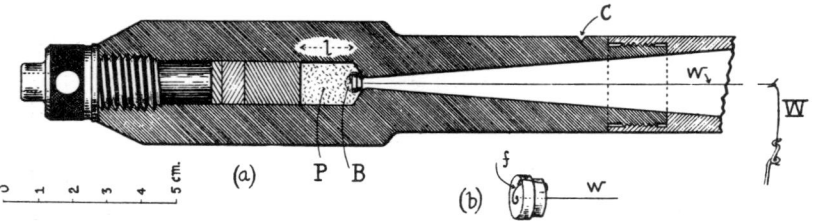

A diagram of Goddard's first rocket – from his essay
'A Method of Reaching Extreme Altitudes' (1919).

depend upon the presence of air for propulsion. A theoretical treatment of the rocket principle shows that, if the velocity of expulsion of the gases were considerably increased and the ratio of propellant material to the entire rocket were increased, a tremendous increase in range would result.'

Goddard continued, 'Experiments repeated *in vacuo* demonstrated that the high velocity of the ejected gases was a real velocity and not merely an effect of reaction against the air. In fact, experiments performed at pressures such as probably exist at an altitude of 30 miles gave velocities even higher than those obtained in air at an atmospheric pressure, the increase in velocity probably being due to the difference in ignition. Results of the experiments indicate also that this velocity could be exceeded, with a modified form of apparatus.'

Next, Goddard outlined his thoughts on Earth-escape rockets.

'For very great altitudes, secondary rockets will be necessary, in order to keep the proportion of propellant to total weight sensibly constant. The most extreme cases will require groups of secondary rockets, which groups are discharged in succession.

'There are, under any circumstances, two possibilities. Either the secondaries may be small, so that each time a secondary rocket, or group of secondaries, is discarded, the total mass is not appreciably changed. (See Sketch A.) Or else a series of as large secondaries as possible may be used in which case the empty casings constitute a considerable fraction of the entire weight at the time the discarding takes place. (See Sketch B.)

'In so far as avoiding difficulties of construction is con-

59

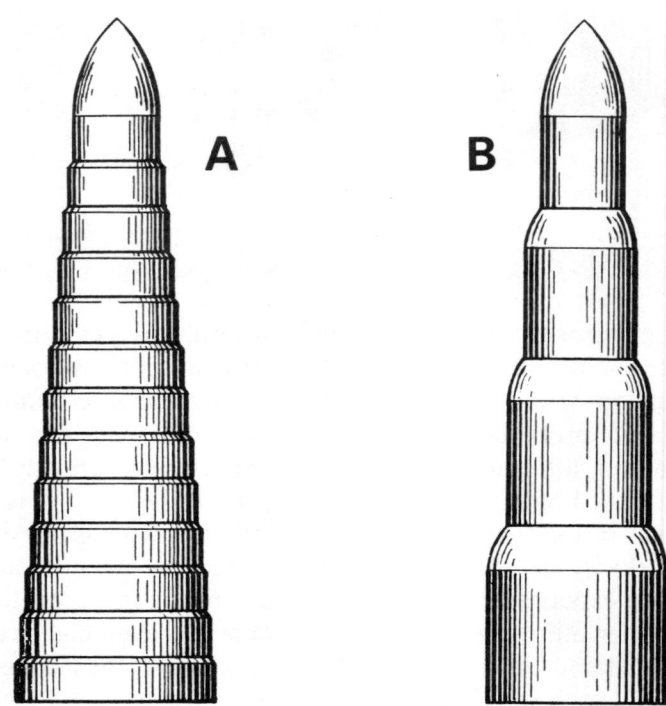

Goddard's multi-stage rocket – another ground-breaking sketch from his work 'A Method of Reaching Extreme Altitudes'.

cerned, the use of a smaller number of larger secondaries is preferable, but they should be long and narrow, as otherwise the air resistance on the nearly empty casings will be greater for the same weight of propellant than would be the case if groups of small secondaries were used, in as compact an arrangement as possible.'

Goddard added to this the highly prophetic statement that only the multi-step principle was, in his opinion, capable of giving the final rocket carrying its useful payload a high terminal or 'characteristic' velocity – generally abbreviated to Vc.

The remainder of the report detailed the various experiments Goddard had carried out – and gave some intriguing hints about those he planned, the most remarkable of which was a moon rocket!

According to Goddard's calculations, he believed that a

rocket with a take-off weight of as little as eight to ten tons would be capable of escaping from the Earth and reaching the moon. This rocket would carry a charge of magnesium powder sufficient when it struck the moon's surface to create a flash that would be easily visible through any telescope exceeding twelve inches in diameter.

Goddard wrote, 'It is of interest to speculate upon the possibility of proving that extreme altitudes had been reached even if they actually were attained. In general, the proving would be a difficult matter. Thus, even if a mass of flash powder, arranged to be ignited automatically after a long interval of time, were projected vertically upward, the light would at best be very faint, and it would be difficult to fore- tell, even approximately, the direction in which it would be most likely to appear.

'The only reliable procedure,' he went on, 'would be to send the smallest mass of flash powder possible to the dark surface of the moon when in conjunction (i.e. the "new" moon), in such a way that it would be ignited on impact. The light would then be visible in a powerful telescope. Further, the larger the aperture of the telescope, the greater would be the ease of seeing the flash, from the fact that a telescope enhances the brightness of point sources, and dims a faint background.'

Goddard revealed that he had conducted a number of experiments with flash powder to ascertain the amount needed for such a project. These had been carried out in small capsules, held in glass tubes, and sealed with rubber stoppers 'in order to reproduce, approximately, the conditions that would obtain on the surface of the moon.' He continued:

'From these experiments it was seen that if this flash powder were exploded on the surface of the moon, distant 220,000 miles, and a telescope of one foot aperture were used – the exit pupil being not greater than the pupil of the eye (e.g. 2 mm) – we should need a mass of flash powder of 2.67 lbs to be just visible and 13.82 lbs or less to be strikingly visible.'

Goddard concluded that a 'total initial mass' of rocket ship plus magnesium powder of '8 or 10 tons would, without doubt, raise sufficient flash for clear visibility.' And he added, 'This plan although of much general interest, is not of obvious importance. There are, however, developments of a general method under discussion, which involve a number of important features not herein mentioned, which could lead to results of much scientific interest. These developments involve many experimental difficulties, to be sure; but they depend upon nothing that is really impossible.'

In fact, history has proved that Goddard was quite wrong to think this idea was 'not of obvious importance'. Writing in the science fiction magazine *Vertex* in February 1974, James Sutherland discussed Goddard's project and pointed out that the exploding charge would not only mark the site of the rocket striking the moon's surface but would also provide some otherwise unobtainable information on the nature of the lunar crust.

Sutherland wrote, 'This quite modest project – and remember that at the time engineers were contemplating a tunnel beneath the English Channel and transatlantic Zeppelin passenger traffic! – was greeted with almost universal derision on two counts. *Everybody* knew that rockets wouldn't function in airless space, and even if they could a pilot had to be aboard to steer.

'But Goddard was totally vindicated on 13 September 1959 when the Soviet robot probe Luna 2 crashed onto the Mare Imbrium, becoming the first man-made object to reach another world. It would be more than a decade before men would follow. Of course, there were vast differences in sophistication between Goddard's crude flashpowder-signal rocket and the radio-equipped space probe, but the basic principles remained fundamentally unaltered. In fact, all unmanned space explorers are designed around a few simple engineering and scientific concepts, many of them centuries old in theory and practice.'

It is, though, a fact that at the time the press and public seized on Goddard's idea, and as Mrs Estha Goddard was later to ruefully recall, 'The newspapers had a field-day over my husband's statement and many called it ridiculous.'

This publicity, which earned Goddard the title of 'Moon

Professor,' completely distorted what he had in mind, and indeed the poor man was soon receiving offers from people who volunteered to *fly* on his rocket to the moon! Among these was a certain Captain Claude R. Collins, the president of the Aviators' Club of Pennsylvania, who wrote to the *New York Times* declaring he was 'ready to fly to the Moon or Mars as long as the professor (Goddard) would provide a $10,000 insurance policy on my life.' A few days later, the paper reported that a young lady from Kansas City, Missouri, was willing to accompany the intrepid Captain Collins!

Understandably, after all this innocently-generated attention, Goddard developed an aversion to any form of publicity that lasted for the rest of his life. Equally, he conducted all his later experiments in great secrecy and even filed away his notes on topics such as 'Methods of Navigating in Space' and 'Manned and Unmanned Missions in Space' in his filing cabinet under the innocuous heading 'Formula for Silvering Mirrors'!

Goddard's space research was temporarily interrupted in 1917 when America entered the First World War – though he was happy enough to answer the call of his country and put his rocket expertise to military use. He directed his attention to missiles for both long and short-range bombardment, and by 10 November 1918 had produced a number of projectiles for demonstration to the military at the Aberdeen Proving Grounds. The largest of these was almost six feet long, weighed 50 lbs, and carried quite a punch, while the smaller, short-range rockets were intended for the infantry to be used against tanks and pillboxes.

The tests were a great success, but before Goddard could take his ideas any further, Germany capitulated and the authorities lost interest in rockets. Interestingly, though, Goddard's prototype missiles were later to be developed into the famous 'Bazookas' – the first hand-weapon to be used by infantrymen against tanks in the Second World War.

By the early 1920s Goddard was concentrating all his rocket proving tests on liquid-propellant motors, using liquid oxygen with such hydrocarbons as petrol, ether and liquefied propane. And by 1923 he had tested his first small rocket in which the liquid oxygen and petrol were fed by pumps to the rocket motor.

This little machine was never in fact launched into free flight, but after a number of further experiments and the perfection of the means of stabilizing and guiding it, a similar rocket employing an oxygen pressure feed was constructed and was ready for an actual test flight on 16 March 1926.

Because Goddard insisted on absolute secrecy, the world was unaware of the historic event which took place on that clear spring morning when, on a farm at Auburn, Massachusetts owned by one of his relatives, the ingenious professor used a blow torch to ignite his fragile-looking ten-foot-high rocket composed of thin sheet steel. There was a flash and a roar, and in an instant the rocket was airborne. It remained in the air for just two and a half seconds, covered a distance of 184 feet, and travelled at an average speed of 64 mph.

The world's very first flight by a liquid-propellant rocket had taken place. It was, as a NASA scientist was later to remark to me, 'the direct ancestor of the Saturn rocket which took man to the moon. Goddard anticipated our space programme by 40 years – and in so doing laid the groundwork which made that programme possible.'

Assisted by his dedicated wife, Estha, Goddard continued his research and on 17 July 1929 fully justified the faith the Smithsonian Institute had shown in him by launching a larger rocket which rose to 90 feet, this time carrying a 'payload' of a barometer, thermometer and a camera, which all parachuted safely back to Earth. Unfortunately, however, because of the veil of secrecy which Goddard placed around his experiments, when the rocket was seen plummeting to the ground in flames, it was reported by curious members of the public as an aeroplane crash! The result was that the state of Massachusetts forbade any more such tests.

The officials at the Smithsonian were, though, impressed by what they saw as the rocket's potential for studying the composition of the upper atmosphere, and came to Goddard's rescue by increasing his grant to $18,000 annually, thereby enabling him to move his operations to the seclusion of the desert area of Roswell in New Mexico. Here he was able to spend the rest of his life experimenting, far from the prying eyes of the media who still hounded him from time to time seeking details of the 'Moon Professor's' latest project.

At Roswell, Goddard secured a brilliant and resourceful assistant called Charlie Mansur, who was later to carry on his mentor's work by helping create the White Sands Proving Grounds where the American space programme first became a reality. As Mansur himself was later to comment, 'At White Sands we had million-dollar recording instruments. But back in the old days at Roswell, we first had to invent our own equipment and then make it by hand out of whatever scraps we could find.' He remembers, too, how Goddard and he were considered in the late twenties to be 'on the lunatic fringe of science'.

Mansur served with Goddard from November 1929 until the great visionary's death in 1945. Their time together in the New Mexico desert was, as he has said, an incredible nightmare of inventiveness, gruelling labour, stubbornness, faith, courage and pure genius. 'There was never so much invention with so little manpower,' Mansur added.

It was a potentially fatal event involving both Mansur and Goddard which ultimately helped the professor finalize his ideas about gyroscopic rocket control – the selfsame gyroscopic control which the Germans were to embody in their V-2 weapons of the Second World War and which later became an essential part of the pioneer Viking rockets of the US space programme. The American historian Lloyd Mallan has recorded this event in these evocative words:

'Charlie Mansur has the dubious distinction of being the only man alive who was ever chased by a rocket. At Roswell, there was a home-made, galvanized silver launching tower sixty feet high, which had been purchased from a mail-order firm. Mansur and Goddard extended its height by 25 feet and painted it with red lead, to protect it from the weather.

'The particular rocket that chased them apparently had gyroscope trouble and gravity took immediate advantage of this. It shot out of the tower to an altitude of 400 feet, then turned around at an angle towards the ground and zipped into a curve. Its direction was now strictly towards Goddard and

Mansur, and it was travelling at about 500 miles an hour. Both of them dropped on their bellies as it whizzed overhead. Mansur ran a few feet first, until he realized that he couldn't compete with the thing!'

But out of failure came success, and Mansur shared Goddard's elation as their rockets climbed higher – to over 7,600 feet by 1935 – and faster – 700 miles per hour. Then in 1940 he launched his most ambitious rocket of all, the 22-foot-long 'P', or pump rocket, named after its centrifugal pump used for forcing propellants into the combustion chamber. With the engine cooled by the gasoline fuel running through copper tubing around the chamber, and the whole mechanics sheathed in a single thin steel covering, it was the forerunner of the space rocket as we know it today both in appearance and performance.

When, the following year, America yet again became embroiled in a world war, Goddard once more put his personal obsession to one side in order to serve his country. Unable to convince the authorities of the potential of his long-range rocket as a weapon, he was instead put to work on developing liquid-fuel jet-assisted take-off units (JATOs) for assisting heavily-laden US Navy seaplanes get into the air in the shortest possible space.

The error of the authorities' way was graphically underlined when the first German V-2 bombs began to fall on Britain. For only the most cursory study of them revealed they were almost identical in design and performance to Goddard's liquid-fuelled rockets. When the professor himself examined the details of the Germans' rockets he felt as if he was looking at his own blueprints. Tragically, he died just a few weeks after the end of the hostilities on 10 August 1945.

According to some people, Dr Robert Goddard died because his frail constitution could not bear up under the hard weather conditions and tension of wartime research. This is *not* a view shared by Charlie Mansur.

'I think he died of a broken heart,' he said in 1956. 'The thing that really hurt him was his own country's failure to recognize him. When the best brains in the country claim that a man doesn't know what he's talking about, the public naturally follows the experts. In my opinion *that* is what shortened his life.'

Mansur was also quick to point out that if America had ignored the great pioneer in their midst, others had not.

'Other countries recognized him,' Mansur said. 'Particularly Germany. The engines of their V-2 rockets were an almost exact copy of Dr Goddard's smaller original. Even their general shape followed the configuration of his engine. And out of the V-2s, of course, have come many advanced and mightier rocket designs. If he had not sacrificed more than a quarter of a century of his life to accomplish controlled rocket flight then White Sands and all the achievements which have followed would still be unwritten history.' (To be fair, America belatedly recognized Goddard's achievement: Estha Goddard was given one million dollars for the use of her husband's numerous rocket patents, while in 1959 NASA chose to name its new operations centre at Greenbelt in Maryland, the Goddard Space Flight Centre.)

The history which Goddard did not live to see was, of course, the conquest of space, the landing of men on the moon, and now the proposed manned mission to Mars. Charlie Mansur has unequivocally called him 'the father of space travel' and certainly the achievement of free flight about the solar system was his greatest dream.

'Others,' Goddard wrote at one stage, 'may speculate about flights to the moon and far-distant Mars. One day we shall certainly reach that cold, mysterious world by super-rocket. But to begin with, we must conquer the first mile into the atmosphere. That will be my aim.'

As we have seen, he did more – far more – than that. For his genius made modern rocketry possible and his inspiration has guided America towards fulfilling that youthful ambition, fired by H. G. Wells, of reaching the planet Mars.

It was an ambition, as we shall next see, that was already being manifested by numerous leading scientists in both America and Russia *long before* man had even reached the moon. . . .

FOUR

'Forward to Mars!'

JUST AS Professor Robert H. Goddard had inspired a few disciples such as Charlie Mansur to carry on his space objectives in America, so, too, in Russia others were following the lead of Konstantin Tsiolkovsky, particularly in the reclusive teacher's enthusiasm for the planet Mars.

In the 1920s and early 30s, as Goddard carried out his experiments in the US, a highly gifted astronautical thinker and engineer named Friedrich Tsander (1887–1933) was enthusiastically promoting and publicizing Tsiolkovsky's work across the length and breadth of Russia. Born in Latvia, Tsander had learned about the research of the old pioneer from Kaluga in 1903 while he was in his last years as an engineering student at Riga Polytechnic Institute. There he had read the first part of 'Exploration of Cosmic Expanse Via Reactive Equipment' and it had, quite literally, changed his life. As Tsander's biographer, L. K. Korneyev has written:

'The whole concept fascinated him . . . Mars, too, captivated his imagination, and in time the planet became an obsession. A slogan he repeated endlessly throughout his life was "Forward to Mars!" '

By all accounts, Tsander dedicated his short and brilliant life to the cause of interplanetary travel, and while still in his twenties had begun actual rocket experiments along the lines which Tsiolkovsky had advocated. In fact, he corresponded with the old man, and was later to edit a collection of his writings. Of Tsander, another biographer, Albert Parry, has written:

'In part following Tsiolkovsky's ideas, in part impelled by his own visions, Tsander designed air rocket engines, liquid-fuelled rockets and rocket planes. His every waking moment was devoted to this wonderful new space science. He brought stars into his very family circle: his daughter he named Astra; his son, Mercury. He was an eccentric, like so many rocket dreamers in Russia, but his schemes proved practical and so he was not laughed down into an insane asylum.

'Unlike the sedentary Tsiolkovsky,' Parry added, 'Tsander incessantly criss-crossed Russia, everywhere bubbling forth his enthusiastic lectures on the coming interplanetary age. Many of today's Soviet rocket experts learned their earliest practical lessons from Tsander.'

One person who was particularly impressed by Tsander was none other than the Communist Party leader, Lenin, who was in the audience at the Provincial Conference of Inventors which he addressed in Moscow in the winter of 1920. In the audience, too, was H. G. Wells, then on his visit to the Soviet Union. He, similarly, could hardly have failed to have been intrigued by Tsander's passion both for space flight and Mars.

The meeting with Lenin deeply impressed Tsander, who later wrote: 'Just before I began my report I was told that Lenin would be present. As I started to talk I was very nervous, but soon I was excitingly arguing to my audience that man's flight to another planet was possible. I recited to my listeners my calculations and design of the future interplanetary ship. Lenin, I saw, was all attention, and this gave me particular strength.

'After my report I was invited to see Lenin. I was very much ill at ease. But he with such simplicity questioned me about my work and my plans for the future that I even took advantage of his time somewhat, and talked to him in great detail about my labours and my dream to build an interplanetary rocket ship. At the end of our conversation, Lenin gave me a firm handshake and promised his support. After this conversation I began to work with yet greater inspiration.'

Although life in Russia at this time was difficult, both economically and politically, nothing could deter Tsander and he was instrumental in the formation of the first Soviet astronautical societies. He was also arguably the first person in Russia to stress the political, not to mention nationalistic, importance of space travel – a fact which was not lost on Lenin and certainly not on Stalin who brought about the elevation of Tsiolkovsky as a 'Socialist Pioneer'.

In October 1924 when he was making a speech at the Great Physics Auditorium at the Institute of Moscow, he was suddenly interrupted by a noisy questioner who demanded, 'But *why* do you want to go to Mars?'

Tsander replied instantly and very definitely. 'Because it has an atmosphere and ability to support life,' he said. 'Mars is also considered a red star and this is the emblem of our great Soviet Army.'

The deafening applause which greeted this remark has continued to echo down through the years of Soviet space history.

The crusading fervour with which Tsander presented the case for space travel, as well as his fascination with Mars, also caught the interest of various Russian writers. In 1923, for instance, Alexei Tolstoy wrote a novel about the planet entitled *Aelita*, which a year later was made into one of the very first Soviet science fiction silent films. Equally successful was A. Bogdanov's *The Red Star*, which ran into six editions, not to mention translations of the works of H. G. Wells, Jules Verne and, later, Edgar Rice Burroughs with his famous Martian series.

But Friedrich Tsander was more than just a writer and campaigner – he was also a practical experimenter, and in 1927 he published his first detailed proposals for an interplanetary ship, as well as showing a model of this at an exhibition in Moscow. In the next six years, Tsander worked on his *Opytnyi Reaktivnyi* or Experimental Jet (OR-1) fuelled by benzine and atmospheric oxygen. Although it has been claimed by the Russians that this was the world's very first liquid-fuel rocket engine, there are no records of it ever having been successfully fired. By the beginning of 1933, however a second rocket engine, the OR-2, powered by benzine and liquid oxygen, was ready for its trials.

Tsander, though, was seriously ill and only ten days after he heard of the successful firing of his rocket he died on March 13, aged just 46. Though he never saw the rocket he had designed actually ascend, he could still write from his deathbed to his friends and followers: 'Forward comrades, and only forward! Hoist our rockets yet higher, higher and higher!'

Like his mentor Tsiolkovsky, Tsander has recently had his achievement commemorated in Russia by the erection of a statue complete with a replica of his rocket, at Kislovodsk, the North Caucasian town where he died.

Even before Tsander's passing, however, another Russian had become deeply involved in the pursuit of interplanetary travel. And following the pioneer's death, he took up the Latvian's mantle and in the fullness of time led the Russian nation into the space age. He was Sergei Korolyev (1907–66) whose life was also bound up by a dual fascination with rockets and the planet Mars. Curiously, too, little was known about this man outside Russia, until his huge funeral in Moscow – as space historian John M. Mansfield has written:

'Korolyev received a State Funeral in 1966 in honour of being the chief designer of the rockets that put Sputnik 1 and Yuri Gagarin into space: Party chief Leonid Brezhnev and President Nikolai Podgorny helped carry his coffin to rest within a few yards of the grave of Stalin. His name was never mentioned in the press, and foreign diplomats were mystified when thousands of Russians braved a snowstorm in Red Square to witness his funeral. If any one man could be credited with being the brains behind Russia's conquest of space, that man would certainly be Sergei Korolyev.'

Three factors are believed to have generated Korolyev's passion for space. As a youngster his abiding interest was in aeroplanes, and it was in the pages of an old copy of the magazine *Vestnik Vozdukhoplavaniya* – Herald of Aeronautics – that he found the second part of Tsiolkovsky's incredible article on rockets. It instantly attracted him to all the old pioneer's writings. Next, in 1924, he saw the movie version of Tolstoy's tale of Mars, *Aelita*. It opened his eyes to the wonders of the distant planet, and at precisely the right moment – for the pages of the Russian newspapers and magazines were full of information about Mars that year because of its close proximity to Earth. Thirdly – and perhaps most importantly –

in 1931 he met Friedrich Tsander. An instant rapport seems to have developed between the two men, and although their approaches to space flight differed, they became firm friends, if all too briefly. While Tsander was undoubtedly the visionary, Korolyev believed the solution to space flight lay in the mating of rockets to the aeroplanes which he had studied since his youth. (It is also possible that he had heard about the work of Robert Goddard, because the furore which surrounded the American's idea of exploding magnesium on the surface of the moon, by the time it reached Moscow in the autumn of 1924, had somehow become erroneously translated to the effect that he had actually *flown* a rocket to the satellite!)

Korolyev was educated at the N. E. Bauman High Technical School in Moscow, where his enthusiasm for flight had been stirred first by learning to fly gliders, and secondly by designing faster and more efficient aircraft under the tutorship of that giant of Russian aviation, Andrian Tupolev. In 1927 he began work in the aircraft industry, rapidly becoming the head of a designer brigade and producing designs both for a new type of aircraft which he called the SK-4 as well as actually building a new kind of glider, the Koktebel. This immediately proved its worth by winning the country's sixth All-Union Glider Competition, staying aloft for the new record time of 4 hours 19 minutes. A partner in this enterprise was another of the now-fabled names of Russian aviation, Sergei Ilyushin.

But it was in the early 1930s that he found his real métier in the pursuit of space travel. In 1930, Korolyev was appointed a senior engineer at the Central Aero-Hydrodynamics Institute in Moscow, and there, a year later, his path and that of Tsander crossed. Despite their different approaches to the problem of interplanetary flight, Korolyev was quick to appreciate the value of Tsander's OR-1 reaction engine although, as I have mentioned, it was never successfully fired. He also undoubtedly helped his by then ailing friend to develop the successful OR-2.

Whether Lenin's promise to Tsander had anything to do with it or not, in 1932 the Soviet authorities officially set up the *Gruppa po Izucheniyu Reaktivnogo Dvizhenia*, Group for the Study of Reaction Motion (Jet Propulsion) or GIRD for short, with Korolyev as its head. Under his auspices, the OR-2 was successfully launched in March 1933. A plan of his own, however, to harness the rocket to a 'flying wing' type of aircraft reached the mock-up stage, but never took flight. Although deeply saddened by the death of his friend Tsander, Korolyev threw himself wholeheartedly into the business of establishing GIRD. He realized now that his idea of fitting a rocket engine to a glider wing was not as elementary as he had imagined, and therefore instituted a series of courses to train skilled technicians in the new art of reaction propulsion. These men were then divided into teams to develop liquid-powered engines and design flight rockets. As such, they were the first 'space teams' – groups now so much a part of both the Russian and American space programmes. By the end of 1933, over 100 people were working for GIRD, and this knowledge makes a nonsense of the often-repeated claims that until several of the German V-2 rockets and a number of the scientists who had worked on them were captured by Russian troops at the end of the Second World War, the Soviets had no interest or involvement in the development of rocketry. It also helps explain *why* the Americans were caught so unawares when Russia launched the first Sputnik.

As John M. Mansfield has written, 'The remarkable growth of interest in rocketry in the Russia of the 1920s was generally unknown in the West until recently. Many Western experts were astonished to find that not only were the Russians working on liquid-fuel propulsion but that they were experimenting with highly sophisticated electric arc rockets, a type of propulsion which is receiving much attention today as being perhaps the next stage of power for space travel.'

It was Korolyev who had first instituted this research, as well as a variety of other far-reaching projects into ramjets, pulsejets and solid propellant projectiles. In 1934 he also produced a thesis – drawing quite considerably on the work of Tsiolkovsky, as it transpired – in which he discussed the problems of high altitude flight and suggested that really large liquid-fuelled rockets would be required to conquer space. He also prophetically pointed out that when manned rocket flight took place, the acceleration and pressurization of the craft would be among the major problems requiring to be solved.

Because of the rapid development of GIRD and the potential of the ideas it was generating, it should not be seen as too surprising that the Soviet military authorities began to take an interest in the wholly civilian organization. Nor, considering the political state of Russia, that shortly thereafter it was decided Korolyev's organization should come under the direct control of the Red Army and its leader, Marshal Tukhachevsky. It was done, though, with Korolyev's cooperation, and resulted in more money for GIRD as well as increased facilities. In return, however, the space flight objectives of the team had to be suppressed in the interest of 'national defence'. Ever since that date, the Soviet space programme has remained a military preserve.

Korolyev and some of his team were, nevertheless, able to continue to develop rockets, and on 17 August 1933 GIRD 09 flew – the first Russian liquid-fuelled rocket to achieve a satisfactory flight. Work went on intermittently thereafter – in between demands from the Army for rocket projectiles for military use. Then, in 1937, came Stalin's infamous 'purges' which swept away many leading political, military and scientific figures. Korolyev himself was arrested along with many others from his organization, but was later allowed to continue his work.

According to G. A. Tokaty, a rocket scientist who later defected to the West, it was these purges which seriously affected the progress being made by people like Korolyev. 'I think that the political arrests and murders of the 1935–40 period, caused much greater damage than is realized abroad. Far too many scientists, technologists and managers were destroyed, humiliated or disheartened. And rocket experts were no exception. Marshal Tukhachevsky was the top governmental spiritual leader of military rocketry. But . . . he was shot in 1937 as a "German spy" and this sparked off a whole chain of disasters. Almost all who worked on a project discussed with or authorized by him, or who were in contact with him – as all leading rocket specialists were – had now to face the danger of being proclaimed an "accomplice of a spy."'

Korolyev, as I have said, escaped the purges, and in May 1939 oversaw the firing of the first Soviet experimental two-stage rocket near Moscow. When the Second World War began, he was employed on designing a flying bomb about which little is known beyond the fact it embodied several of the principles later employed by the German scientists in their similar offensive weapon.

Surviving the war as well as the purges, Korolyev emerged unscathed to become one of the leading rocket-builders of the Sputnik era. He was appointed 'Chief Spacecraft Designer' and directed the construction and launching of the first Sputniks as well as the later manned Vostoks.

It was a fitting honour for a man who had seen the dreams of his mentors Tsiolkovsky and Tsander become reality. And it was perhaps those two pioneers as much as Korolyev himself that the silent masses standing in Moscow unconsciously honoured at his funeral. For by now the dream they all shared of reaching other planets was at last within grasp.

□

In Frank H. Winter's book, *Prelude to the Space Age* (1983), in which he recounts the histories of the various small experimental 'Rocket Societies' which suddenly appeared in both Europe and America in the years between the two world wars, he has this most revealing comment to make.

'There is a striking parallel between the story of the Soviet pioneers and those of the German Rocket Society, the *Verein für Raumschiffahrt*, or VfR,' he writes. 'In both countries the men started in the early 1930s as idealists pursuing rocketry as an avocation with a vision towards an eventual space or stratospheric vehicle. In both countries one of their number was to emerge as exceptional leaders – Korolyev in the USSR, and Wernher von Braun in Germany. By 1935 the Armies of both nations clearly diverted the idealistic path towards militaristic aims. At the opening of the Space Age, von Braun directed the United States' Explorer 1 mission and afterwards the Apollo programme. In the Soviet Union, Korolyev had become the Chief Spacecraft Designer who directed the

Sputnik and Vostok projects. In both countries the respective design and launch teams also included several other 1930s rocketeers. The legacies of the VfR, GIRD and other such societies are manifest. They served as experimental test beds and schools for far greater things to come.'

Mention of Wernher von Braun brings us to the last of the great rocket scientists who was as equally fascinated and driven to reach Mars as any of the others. Indeed, he has played arguably the greatest part in enabling man to stand on the threshold of this achievement, for his V-2 rocket is now rightly seen as 'the blueprint for travels to Mars', a term given to it by historian Brian Ford.

Von Braun (1912–1977) was yet another young man whose fascination with space flight had been inspired by the books of Jules Verne and H. G. Wells, and whose driving ambition to reach Mars had been fired by a particularly vivid sighting of the red planet which he had had when looking through a telescope while standing in his garden, a wide-eyed twelve-year-old, in 1924. The reader will perhaps recall that the planet was in particularly close proximity to the Earth that year.

As the son of a Weimar Republic agricultural minister, von Braun grew up in some affluence: well-educated, sophisticated in manner and prodigiously imaginative, he discovered the VfR Rocket Society not long after his sighting of Mars, and by his late teens had become a leading member. His inspiration was undoubtedly Professor Hermann Oberth, the man regarded as the 'father of German rocket development' and the author of *Die Rakete zu den Planetenraumen* (The Rocket into Planetary Space) a slim, 92-page volume published in 1923 which laid down many principles concerning space navigation, guidance and landing, as well as the outlines for a two-stage unmanned rocket and an egg-shaped two-passenger spaceship – all of which have subsequently proved to be completely viable. Oberth had, in fact, been much impressed by what he had heard of Robert Goddard's work, and though he wrote to the American pioneer for a copy of his 1919 publication, there is no question that he copied his contemporaries' ideas.

By the time he was 18, von Braun was an enthusiastic assistant at Oberth's rocket experiments. Although he was then apprenticed at the Borsig engineering works that specialized in railways, his mind was clearly on more interplanetary things, in particular the professor's *Kegelduse* or Cone-jet rocket motor, named after the shape of its steel and copper combustion chamber. After a number of problems, the *Kegelduse* was successfully fired before a group of observers on 23 July 1930, running for 90 seconds and delivering a thrust of 15.4 lbs. According to the VfR, it was the first officially certified rocket to have run anywhere in the world. (Goddard's tests, were, as I have noted, carried out in the strictest secrecy.)

Inspired by this success, von Braun and the other VfR members next began to work on the Mirak or 'Minimum Rocket', a small, four-foot-long projectile which used liquid oxygen to cool the copper combustion chamber where the fuels mixed and fired. Unhappily, two static tests in 1930 and 1931 both ended in ruinous explosions.

Although the members of the VfR were not to be deterred from their objectives, they were naturally concerned about the risks entailed by the experiments. Clearly, they needed their own 'rocket-flying field' as one member put it. A piece of suitable land was searched for and eventually found on the outskirts of Berlin. It had formerly been an ammunition dump, and though unprepossessing in appearance was ideally suited to rocket tests. Some old buildings were refurbished and turned into workshops and living quarters, while dug-outs and tunnels were built in which the experiments could be made. The *Raketenflugplatz* was the first site of its kind, and several historians have pointed out that it was here that the tests which eventually enabled the Russians and Americans to conquer space were begun.

The lack of facilities were more than overcome by the enthusiasm of those who came to work for the VfR, as von Braun later recalled. 'Our labour force cost nothing,' he said, 'by reason of the then prevailing general unemployment. Many a draughtsman, electrician, sheet-worker and mechanic was only too happy to take up residence rent-free in one of our buildings and to maintain his skill at his trade. Soon there were some fifteen craftsmen living in our refurbished buildings and working eagerly on the tasks we set them.'

The dedication of all these people was rewarded when a new rocket, the Repulsor, was developed in the late spring of 1931. This time the motor was fitted with aluminium walls and cooled by water instead of oxygen. In three tests, the rocket reached heights of 60 feet, then 200 and finally 600 feet. By 1932, an improved version of the Repulsor could fly more than 3 miles horizontally, and almost 1.5 kilometres (0.9 miles) high. The VfR next added a long stick to the Repulsor for greater stability, and this version shot over 3,000 feet into the air before parachuting safely back down to the ground.

The economic plight into which Germany had now fallen (over six million were unemployed by 1932) also brought the VfR to crisis point. The saviours – if such they can be described – were the German Army who like their Russian counterparts had been taking a cautious interest in the activities at the *Raketenflugplatz* for some months. As a result of the Treaty of Versailles, the country could not manufacture heavy artillery. But nothing had been said about rockets. After a successful demonstration of one of the Repulsor rockets at the Army's testing grounds at Kummersdorf, invitations were offered to several of the VfR workers to continue their research under the patronage of the Army.

Foremost among these was Wernher von Braun who understandably found the offer to work on a new rocket programme, generously funded and with far greater facilities, quite irresistible. It was a decision, one should be quite clear, that was totally without political motive. In all that followed, von Braun was driven by the desire to achieve space flight, *not* to serve Hitler and his mad desire for world conquest. He left the *Raketenflugplatz* for the Army in November 1932, and by January 1934 the once flourishing VfR was no more than a memory – the Berlin testing grounds which had harboured its greatest dreams once again returned to being just an ammunition dump.

Von Braun's subsequent work for the Army, following the establishment in 1937 of the secret rocket site at Peenemunde on a Baltic island of which he was made technical director has, of course, already been told in such detail as to require no more than outlining here. Suffice it to say that he was joined in this work which was to culminate in the war's most fearsome weapon, the V-2, by a number of former VfR members, all of them labouring in the same spirit that had motivated them before.

'I was sure that the site of the *Raketenflugplatz* was utterly inadequate even to commence the vast experimental programme which must be the precursor of successful rocket flight,' von Braun wrote later. 'It seemed that the funds and facilities of the Army were the only practical approach to space travel. At that time none of us had any idea of where the work would lead us or any thoughts of the havoc which rockets would eventually wreak as weapons of war.'

More proof of von Braun's motivations – if such are still required – may be gained from the fact that he was actually thrown into prison by Himmler for refusing to transfer the allegiance of his Peenemunde staff to the Gestapo. It required the personal intervention of Hitler to get him released after two weeks of uncomfortable incarceration.

Von Braun had begun designing his prototype of the V-2, which was then known as the 'Assembly-4' or A-4, in 1935. It had a length of 25 feet, a thrust of 3,300 lbs and employed vanes which operated in the exhaust gases to maintain a preset flight pattern. After several years of experimentation, this ultimately hugely influential rocket had been perfected. It then stood 46 foot high, had a thrust of 59,500 lbs, and an estimated range of 200 miles. It was not only the world's first long-range ballistic missile, but to von Braun and his colleagues was also an 'embryo spaceship'. We are fortunate in having the designer's own description of this moment of history.

'The first experimental A-4 was fired from the Peenemunde site in the spring of 1942,' he recalled. 'It rose with a faint swaying motion, until it was lost to view in the clouds. The noise of the motor could still be heard, then it stopped abruptly! A few seconds later, the A-4 fell through the clouds, minus its fins; tumbling end over end, it fell just off shore and sank after a tremendous explosion. The second A-4 also went astray after climbing several miles, but the third attempt on October 3 proved completely successful. The rocket was fired through an overcast sky, but the motor was heard for the correct period of time, and later fishermen reported a mysterious 'plane crash' in the target area in eastern Pomerania.'

German Army leaders, delighted at the success of the A-4, ordered that it now be adapted to carry a high-explosive warhead for use against Allied targets. It was also redesignated the *Vergeltungswaffe-2* (Vengeance-weapon) or V-2.

Much as he would have preferred to have continued developing the rocket for space flight, von Braun – aided by his old mentor, Professor Oberth, who had now joined him at Peenemunde – reluctantly buckled down to the task. In September 1944 the first of some 3,000 V-2s were launched against targets such as Antwerp and London – and only the fact that they became operational when the war was all but lost undoubtedly prevented the rockets from delivering victory to Hitler.

Von Braun, though, continued to look beyond the war, and clearly sensed where the ultimate destination of his creation lay, as this quote from another of his essays reveals.

'One day at Peenemunde I fired a V-2 on a clear evening 15 minutes after the sun had set. The stars were already coming out, and as the great rocket climbed upward the flame of its exhaust diminished to a shining pinpoint and disappeared. Then the rocket broke into sunlight above the shadow of the Earth and gleamed, brilliantly visible, against the darkening sky. At that moment, all space seemed within its reach.'

General Walter Dornberger, to whom von Braun was responsible, was himself caught up in the younger man's enthusiasm for space by this time, and though aware of his primary military objective, could similarly foresee the future.

'That third day of October 1942 was the first of a new era in transportation, that of space travel,' he said later. 'So long as the war lasts, our most urgent task can only be the rapid perfecting of the rocket as a weapon. Then the first thing will be to find a safe means of landing after the journey through space.'

The defeat of Germany in 1945 might well, of course, have

seen the end of von Braun's dream had not he and a number of his fellow scientists as well as a considerable number of the actual V-2s survived. Von Braun and Dornberger surrendered to the Allies, and after rigorous interrogation were allowed to emigrate to America. There, aided by the plans he had already been drawing up for more advanced rockets, plus enough pieces for nearly 80 of the V-2s which had been shipped across the Atlantic, this genius of rocketry evolved the now-famous programme for experimentation and testing which climaxed in 1969 with Project Apollo which put the first man on the moon.

The Russians, in the meantime, had also recruited a number of the Peenemunde scientists and salvaged a number of the V-2s to augment the plans of Korolyev for space conquest. The expertise these Germans had gained in the V-2 programme – conservatively estimated to have cost their nation over one hundred million pounds – was doubtless a crucial factor in enabling the Soviets to surprise the world with the launch of the first artificial satellite, Sputnik 1, just fifteen years later on 4 October 1957; this to be followed by the first man in space, Yuri Gagarin, who made his historic single orbit of the Earth on 12 April 1961.

There seems little doubt, with hindsight, that Wernher von Braun was far enough advanced in *his* planning to have been able to precede both these dates had he been given the go-ahead. As it was, in 1960 he became director of NASA's George C. Marshall Space Flight Centre in Alabama, and there masterminded the even more impressive manned landing on the moon. Even at this moment of triumph, however, his thoughts were on still greater goals: that persistent dream of reaching Mars was in the forefront of his mind.

Von Braun had, in fact, been mulling over ways of reaching the red planet ever since those early days at the *Raketenflugplatz*, but it was not until after the war that these were committed to paper in a work he called *Das Marsprojekt*. Though some of the ideas are undoubtedly impracticable, there are others which, as we shall see later, clearly foreshadow what is now considered the most practicable method of making a manned landing on the planet.

In 'The Mars Project' von Braun suggested building a fleet of ten spaceships to carry a team of seventy men from the Earth to Mars. These ships would have to be built in Earth orbit about 1,075 miles above the surface of the planet, he said, and to facilitate this he proposed a series of 6,400-ton three-step rockets capable of carrying payloads of 39 tons each.

On departing from the Earth's orbit the spaceships would each weigh 3,720 tons, and with 200 tons' thrust, the firing time required to attain the necessary transfer velocity would be no less than 66 minutes. Once the fleet of ten ships had reached Mars, he said, they would orbit the planet, releasing three 200-ton winged rocket planes – similar to the Earth return vehicle – carrying in all some 50 astronauts who would then land on the planet's surface and perform a detailed survey and exploration.

Von Braun explained that one of these landing rockets would carry the supplies for the expedition and would be abandoned after it had served its purpose. After a stay on the planet of 400 days, he said, the party would return to the orbit-to-orbit vehicles and thereafter begin the return journey to Earth, the round trip taking a total of 2 years and 239 days.

'To establish this project,' he concluded, '46 of the big 6,400-ton Earth-to-orbit vehicles would be needed, each making on an average 21 flights, giving a total of 950 flights over a period of eight months. For these flights no less than 5,320,000 tons of a nitric acid-hydrazine propellant would be needed whilst the propellant required for the interplanetary voyages itself would be no less than 36,600 tons.'

After writing those words, von Braun continued to ponder on the challenge of Mars and also design still larger and more complex rockets – those selfsame rockets, the Redstones, Vikings and Saturns, which have afforded the Americans such success in space. And in 1969, immediately after the moon triumph, he set the tone for US endeavour in the years leading up to the end of the century with yet another remarkably prophetic statement.

'My personal guess,' he said, 'is that for a number of years we'll continue to explore the planets with unmanned equipment to get as much information on them as possible. We will not limit this solely to the nearer planets like Venus and Mars, but also take an interest in the outer planets, Jupiter and Saturn. We know so little about them.

'Man will then probably follow to a few of these planets wherever he finds the environment interesting enough, and he may use nuclear rockets to get there. That, I believe, is anywhere between one and two decades away – and foremost will be a manned flight to Mars!'

These were thoughts – as we shall see – that others were sharing, too. . . .   □

In Russia, scientists spurred on by the examples of Tsiolkovsky, Tsander and Korolyev had been earnestly thinking about trips to both the Moon and Mars for years before the first Sputnik was launched in 1957. As the Western commentator Albert Parry wrote in 1954, 'The dearest wish of today's Soviet astrophysicists and rocket-makers is to reach, within their lifetimes, a level of their science where spaceships would soar to – and return from – journeys to other planets. . . . And as this dream has coincided with the Cold War era, so it has become a political race between East and West, between the Soviet Union and the United States.'

Foremost among these Russians was Professor Ario A. Shternfeld of the Academy of Sciences, who confirmed Parry's statement in an article published in *Pravda* in May 1954 in which he speculated about a Soviet liquid-fuelled rocket which would reach the moon 'in just a few days' and Mars 'in a matter of weeks'. If Shternfeld was rather vague in this essay, for whatever reason, he was much more precise – indeed remarkably so for a Russian at this time – in a second essay, 'A Trip to Mars' published a year later.

'A flight to Mars will be of great interest,' he wrote. 'During the past three centuries this planet has attracted particular attention on the part of astronomers and other scientists because of its proximity to the Earth and similar natural conditions. The experts are no longer satisfied with studying the surface of Mars from images that appear tiny even through the largest telescopes.'

Shternfeld said that he believed any trip to Mars with a landing on its surface would be preceded by reconnaissance flights around the planet. 'For this purpose,' he continued, 'spaceships will temporarily become artificial satellites of Mars. However, landing and take-off will be extremely difficult in the early stages of space-travel – the chief trouble being that fuel for the return journey will have to be brought from the Earth.'

With keen foresight, the Russian Academician correctly predicted that a detailed examination of the Martian planet would first be required to select suitable areas for landing, as well as a careful study of the structure and composition of the atmosphere to establish whether it could be used for 'slowing down the spaceship' and also if it would be hostile to human beings.

Next he boldly revealed what was obviously the Academy's thinking about observation flights around Mars. These, he said, would be made along different trajectories: 'the duration of the trip and the initial speed of the spaceship depending on the trajectory chosen.'

'Let us consider,' Shternfeld wrote, 'a trajectory involving a two-year journey. The rocket will start from an interplanetary station at midnight local time when the centres of the Earth, the sun and the station are in a straight line. This is the most appropriate moment because the direction of the station's motion and that of the starting rocket will then coincide. Taking advantage of the speed of the spaceship, the rocket will take off with the lowest speed of 4.3 kilometres per second, whereas a speed 12.3 kilometres per second would be needed were it to leave from the Earth.

'A rocket weighing ten tons with an exhaust velocity of four kilometres per second must have 19.6 tons of fuel on board if it is to take off from the interplanetary station, and 216 tons if it is to be launched from the Earth. The rocket's speed will constantly change as it flies in interplanetary space, being at its highest during take-off, and gradually decreasing as the rocket recedes from the Earth's orbit.'

Shternfeld went on, 'Having approached Mars, the rocket

will by-pass it at a certain distance and fly off into outer space. During the flight around Mars the traveller will be able to photograph the entire surface of the planet, owing to its rotation on its axis. One year after the take-off the spaceship will reach the farthest point of its trajectory, at a distance of 2,175 light-years from Earth. After passing this point the spaceship will once more approach the Martian orbit at an increased speed. But this time it will not meet the planet. The elliptical trajectory of the flight having closed, the spaceship will return to the Earth at the speed at which it took off.'

The Russian Academician also made the interesting suggestion that 'more powerful rockets' could land on the Martian moons, Phobos and Deimos, and there conduct research about the mother planet. Lastly, and most interesting of all, he considered a *manned* landing on Mars.

An economical trajectory in terms of fuel and time for the spaceship's flight was of paramount importance, he said, and continued, 'Let us assume that a journey to Mars is launched along the shortest and most direct route. In that case it would be completed in 85 days, but a speed of not less than 39 kilometres per second would have to be attained for this purpose. It is obvious that such a route would be extremely uneconomical. On the other hand, a spaceship flying along a semi-elliptical trajectory would have to take off from the Earth at a minimum speed. When coming in for a landing on Mars its speed would also be its slowest.'

Shternfeld next came to the heart of the matter. 'It has been pointed out that the spaceship cannot take off from Earth at any moment unless it follows a straight course,' he said. 'If the rocket is to meet Mars when it reaches its orbit, Mars must be in a particular position in relation to the Earth, a position that occurs once every 780 days on the average.

'A journey to Mars along a semi-elliptical trajectory would take 259 days. Before starting on their way back along the same trajectory, the space-travellers would have to wait 454 days for the two planets to stand in the correct relation to each other. And a spaceship flying to the planet on such a trajectory would have to attain a speed of 11.6 kilometres per second at take-off.

'However,' concluded Shternfeld, 'it is doubtful whether would-be space-travellers would choose to fly on such a long route. In order to cut transit time they would probably increase take-off speed and fly along a parabolic trajectory. In that case their journey would last 70 days, provided the spaceship had an initial speed of 16.7 kilometres per second. Consequently, if the initial speed is increased by 1.4 times, the duration of their voyage will be reduced by a factor of 3.7. That is a remarkable feature of navigation in space.'

Remarkable was also an adjective that could be applied to the article itself, for it clearly alerted those in the West that the Russians had their eyes firmly fixed on Mars as a prime objective. The Americans, though, did not immediately respond to the challenge – but when they *did* in the autumn of 1961 it was with the forthrightness which has since become their hallmark in space.

'USA TO LAND ON MARS IN 1972' was the announcement made at a space conference in New York on October 9 and headlined around the world the following morning. Britain's leading science expert Chapman Pincher wrote in the *Daily Express*:

'A precise date for the first United States manned landing on Mars was given to the space-flight conference in New York today. It is 22 May 1972.

'Three cosmonauts will touch down then on the red planet after a journey of 34,000,000 miles, lasting seven months. After 100 days of exploration and looking for Martian life they will take off for home, getting back on December 15.

'Scientists of the Douglas Aircraft Company say the trip will be made in an atom-powered spaceship like an ice-cream cone 157 feet long.

'The United States' lead in nuclear propulsion – in which Britain, again, is doing no work – offers the Americans their best chance of beating the Russians to Mars and of colonizing the moon.'

If the landing date on Mars has subsequently proved to be highly optimistic, it nonetheless demonstrated the American determination: a determination the Russians quickly countered.

Three months later, Yuri S. Khlebtsevich, the chairman of the Soviet Technical Committee on Guidance of Rockets, announced what he called his 'L-V-M Timetable' in these rather grandiose words:

'There is no doubt that in the very next few years our brains, willpower and toil will prove to be stronger than the Earth's gravity, and that in the black, starry sky above there will rush our Soviet rockets filling with pride and joy millions of hearts of the Earth's people.'

Khlebtsevich's 'timetable' proved to be a schedule of Lunar-Venus-Mars flights, showing approximately when, by what year if not month, the Russians would reach the three named solar bodies. Despite the title, the author several times listed Mars as being reachable *before* Venus, and predicted a landing as early as the next year, 1962, but certainly by 1971 ahead of the Americans! A three-stage rocket complete with fuel and having a total weight of 6,000 tons would be needed for such a trip, he said, and would require 256 days in each direction. About how near to completion such a spaceship was, Khlebtsevich was suitably silent!

For once there was a split in the Russian ranks, for several of the elder statesmen of science were evidently not prepared to go along with such a bold claim. Academician Leonid Sedov, for one, said that while it was perfectly realistic to talk of manned landings on Mars, 'to predict exact dates for astronautics is a dangerous thing.' He did, though, think such flights would occur this century and that, 'somewhere in the Soviet Union there is a youth today who, though he may be unaware of it, will be the first to land on Mars.'

A similar optimism was also aired by the triumphant Lieutenant-Colonel Yuri Gagarin after he had become the world's first spaceman. Speaking in August 1962, he said he saw a manned landing on the moon as coming in the next decade, and that this would be only the first step towards Mars.

'I hope,' he said, 'that long before 1981 the first astronomical observatory and the first cosmodrome for flights to Venus or Mars will have appeared on the moon.'

In the interim, he visualized unmanned transport rockets landing at convenient spots on the moon's surface to carry out extensive programmes of research. These in time would then deliver everything necessary for people to live there – food, rocket fuel, mining machinery and prefabricated parts to be used in building living accommodation for the colonists.

Gagarin said that it was the opinion of scientists in Russia that by understanding the nature of other planets – Mars, in particular – it would also be possible to understand our own world better.

'Just as the Sputniks have enabled us to accurately determine the shape, composition and size of our Earth, so a rocket to Mars might well answer other equally important questions. For if a rocket could establish that the planet – whose origins are closer to those of Earth – has a magnetic field, then the nature of the Earth's core would also be definitely known.'

Such possibilities could only enhance the attraction of reaching Mars, Gagarin claimed. Continuing, he said, 'Some people may accuse me of being too modest in my dreams, and say that 20 years from now thousands of tourists will be able to go not only to the moon, Venus and Mars, but even to Mercury and Pluto. But that is certain to happen in time, if not for an epoch or two.'

Although the Russian talk of colonies on the moon was dismissed by some experts like the physicist Sir John Cockcroft as 'no more than fairy tales to justify their enormous space efforts', others such as the distinguished British astronomer and director of Jodrell Bank Experimental Station, Sir Bernard Lovell, were convinced it was just one more indication of the Soviets' intention to reach Mars and the other planets – and well ahead of the Americans.

Sir Bernard voiced his conviction on 18 June 1963 when the Russians pulled off yet another coup by putting the first woman into space. After Lieutenant Valentina Tereshkova had returned to Earth having spent two days in orbit in Vostok VI, he was quoted by *The Times* newspaper as saying that he believed this to be 'part of a long-term plan to colonize another part of the solar system in about 25 years time.'

The Soviets' main objective, Sir Bernard said, was Mars, 'which has the least hostile environment'. The problems of living there, he thought, 'would be no greater than living in a jet airliner.'

The Times report also added, 'Sir Bernard said it would be important for such a programme to establish the effects of weightlessness and radiation damage on both the female and male. Sending a woman into space was certainly not a "stunt" he added.'

The evidence was now there for all to see. The Russians were gearing themselves up for a landing on the moon and then planned a similar mission to Mars. The Americans, though they lagged behind at the moment, had the selfsame objectives.

But in an inspiring address in April 1963, President John Kennedy set the tone for the next decade.

'I think that having made the decision that we should make a major effort to be first in space,' he said, 'I believe we should continue to do so.'

And his words were emphasized by Vice-President Lyndon Johnson who added, 'Slowing down now might bring us one of these days face to face with the shattering knowledge that we had permanently assigned ourselves to second place.'

It was a call to action that was, of course, to be answered by NASA in the most spectacular fashion with the Apollo moon landing. But even before this occurred, there were voices insisting that a moon landing would not establish the ascendancy of one of the superpowers over the other, any more than the Sputnik or Gagarin flights had done previously. The acerbic *New Republic* put the matter bluntly.

'The Moon Race will not be the last,' it declared in a thundering editorial in April 1963, 'for the loser will almost certainly challenge the winner to another race. This time to Mars, and the contest will continue.'

The race for Mars was now clearly beginning to gain real momentum.

FIVE

Touchdown on Another World

THE CREDIT for the very first attempt to make contact with Mars clearly belongs to the Russians, although it was to be the Americans who first successfully orbited the planet and also obtained the first close-up pictures of the red world. As one might expect, this lap of the race for Mars proved enthralling, as well as a little intriguing, as the two superpowers jostled for supremacy with their pioneer space probes.

It was Sergei Korolyev's team in Russia who attempted the first Mars probe in mid-October 1960. Korolyev's fascination with the planet had made him well aware of its elliptical planetary progress around the sun which every two and a half years brings it into relative alignment with the Earth. At these times, he knew, a kind of 'launch window' opened which would allow a rocket to reach it with the minimum expenditure of fuel. When, in that October of 1960, the 'window' opened, it was just too tempting an opportunity for the great Mars enthusiast to resist.

Permission had, in fact, already been given by Premier Khrushchev for some kind of 'spectacular space shot' to be undertaken as a publicity-winner which would coincide with a visit he was making to the United Nations in New York. Rumours had been encouraged that the spacecraft being used might even be *manned*, as a number of prototype manned spacecraft were known to have been tested by the Russians earlier in the year.

As it transpired, nothing happened at all, and it was some years later before any information was forthcoming. It seemed an attempt *had* been made to launch not one, but two probes to observe Mars – but there had been a disastrous explosion at the launching site in Kazakhstan, and several of the leading space scientists had been killed. With typical Russian thoroughness, though, the whole incident was covered up as if it had never happened.

It was only when the memoirs of the former Russian spy, Oleg Penkovsky, were published in 1965 that the facts came to light. The first of the rockets had apparently been launched on October 10, and although Penkovsky knew none of the details of the failure, he believed it had occurred high in the sky over Kazakhstan. A US specialist who examined this account believed the disaster may well have been caused by the failure of the turbopumps in the rocket's upper stage which would have come into operation when it was about ninety miles up.

The second Russian try for Mars was made four days later on October 14, and may well have been undertaken in some desperation to please Khrushchev, then doubtless growing agitated in New York. This probe, however, failed even to leave the launch site, toppling over as it ignited and then exploding in flames to cause grievous damage and the deaths of a con siderable number of scientists and site workers.

Despite these failures, the Russians were not to be deterred, and, two years later, successfully launched a pair of space probes towards Mars. The first of these, appropriately called Mars 1, was dispatched on 1 November 1962, and twelve months later was flying by the planet. For a time all seemed well as the little craft began to relay data back to Earth.

Then, suddenly and inexplicably, faults developed inside the probe and radio contact was lost. At the Kazakhstan space centre, the scientists worked desperately to regain contact, but without success. When Mars 1 was last heard from it was spinning aimlessly some 60,000,000 miles from Earth.

The second craft, Mars 2, was rather more successful. It, too, reached Mars and then commenced the second part of its mission – to land on the surface of the planet. One can only guess at the emotions that must have gripped Korolyev back on Earth as the probe neared touchdown.

Once again, all seemed to be going well as the tiny craft edged towards the rocky terrain. Then came the clear indication the vehicle had landed and for twenty magic seconds signals were received for the very first time from the surface of another world. But as suddenly as the transmission had begun, it ceased – and nothing more was ever heard from Mars 2.

The Russians were, for the first time, prepared to admit to this mission – for it had certainly been a partial triumph. The cause of the failure of the craft they reasonably attributed to a raging dust storm with winds in excess of 200 miles per hour – which they had established was taking place at the time. Had the weather not been against Mars 2, it seems highly likely the craft would have sent back to Earth the very first pictures of the surface as well as the results of some simple chemical analyses of the soil and atmosphere.

As it was, this success was to be achieved two years later by the Americans who, on a landmark day in the summer of 1965, at last brought the red planet into intimate close focus. After centuries of only distant views from Earth-bound telescopes, the little probe called Mariner 4 climaxed its journey across the yawning millions of miles of space and flew by Mars, just 7,800 miles from the surface, to send mankind their first real views of the planet.

It was a moment that fulfilled many of the dreams of those pioneers – both American and Russian. (The Russians, incidentally, though they trailed the Americans at this juncture, dispatched no less than four Mars landers between 10 February and 12 March 1973, which all alighted successfully in 1974 and returned pictures and data similar to that subsequently collected by the Americans and described here.)

Although the space probe Mariner sent back just 21 images as it passed Mars on its interplanetary mission which had begun from the Kennedy Space Center at Cape Canaveral many months before, they were at least 150 times more detailed than any telescope had ever obtained. And though the pictures covered less than one per cent of the planet's surface, they radically affected man's view of his neighbour in the solar system. They also virtually solved the enigma of the mysterious *canali.*

Among those delighted observers of the Mariner photographs was the famous astronomer, Carl Sagan, who commented later that although they were taken in a season of minimal visibility, 'they nevertheless uncovered a variety of apparently rectilinear features.' He was also fascinated by the craters which the pictures revealed – ranging in size from two miles to over seven miles across.

Summing up, Sagan said, 'The radar evidence and Mariner 4 photography suggests that the Martian "canals" are actually ridge systems or associated mountain chains in analogy to similar features in the terrestrial ocean basins. In such an interpretation much of the observed geometry and variability of the canals can be understood. Future radar observations as well as an improved understanding of terrestrial submarine relief would be useful in further testing these views.'

There was a general consensus of opinion among the scientists that Mars appeared to be a rather barren world not unlike the moon. There was also little evidence from Mariner 4's report that the red planet had any of the romance about it that the storytellers of old and the astronomers like Schiaparelli and Lowell had visualized.

This view was, if anything, emphasized when the next two probes, Mariners 6 and 7, again flew by the planet in 1969. Although both passed much closer to the planet's surface – just a little over 2,175 miles – they sent back images of a terrain covered by craters, although there was some evidence of erosion having taken place. Only when a few pictures of what was apparently a cap of frost at the south pole were spotted among Mariner 7's 126 images, were any doubts expressed among the NASA scientists about their initial verdict.

With the benefit of hindsight, J. F. McCauley, a leading member of the Mariner team, was able in 1976 to explain *why* the planet's geology had at first seemed so unpromising.

Speaking after the success of the next mission in the series, Mariner 9, which generated what has since been called a 'bonanza of Martian information', Mr McCauley said: 'Soon it became apparent that almost all generalizations about Mars

93

derived from Mariners 4, 6 and 7 would have to be modified or abandoned. The participants in earlier fly-by missions had been victim of an unfortunate happenstance in timing. Each earlier spacecraft (except in part for Mariner 7 which had returned startling pictures of the south polar regions) had chanced to fly by the most lunar-like parts of the surface, returning pictures of what we now believe to be primitive, crater areas. Given a difference of as little as six hours in arrival times of any of these earlier spacecraft (each of which had spent many months in transit), an entirely different view of Mars would have resulted. It was almost as if spacecraft from some other civilization had flown by Earth and chanced to return pictures only of its oceans.'

The Mariner 9 project had promised to be an exceptional one from its very inception – though not one without its fair share of dramas. The mission was initiated in 1968 with the intention of putting twin probes, Mariners 8 and 9, into actual orbit around Mars to carry out reconnaissance and map the two halves of the planet. Unhappily, on 9 May 1971, Mariner 8 was lost in a launch-vehicle failure, and its twin had to be speedily programmed to carry out the entire mission on its own.

Safely launched from the Kennedy Space Center on May 30, Mariner 9 journeyed for 157 days of interplanetary flight before reaching Mars on November 14 – thereafter becoming the first man-made object ever to orbit another planet.

But the drama was not yet over. A huge planet-wide dust storm, like the one which bedevilled the Russians, had blown up and was completely veiling the surface. Although this obviously forced a postponement of the mapping mission for many weeks, it did provide the scientists on Earth with an excellent opportunity to study the disturbance. When the weather finally cleared, Mariner 9 began a period of intense study of Mars which yielded an unprecedented haul of information, as a delighted NASA Administrator, James C. Fletcher, later recalled.

'After the storm abated, Mariner 9 set about a mapping and scientific reconnaissance of exceptional quality and value,' he said. 'It photographed virtually the whole surface of the planet, sent more than 7,000 images back to Earth, and relayed a total of more than 30 billion bits of information, an amount equivalent to 36 times the entire text of the *Encyclopedia Britannica*. This is incomparably more than has been received from all earlier planetary missions put together.'

Of equal importance to quite a number of people was the thought that with such close-up, high definition photographs of the surface, the mystery of the Martian canals could be solved once and for all. Right off, it was clear they were definitely *not* waterways, *nor* were they vegetation. Almost as emphatically, surface cracks and ridges could be ruled out. So where did that leave the men from NASA?

Carl Sagan, to the forefront again, reported that it was certainly possible now to discount the past reports of canals and relegate them as 'monuments to the imprecision of the human eye labouring under difficult seeing conditions.'

But, he asked, did the story *really* end there? 'Probably so,' he said, 'but the consistency of the diverse maps leaves a lingering unease. Maybe the facts that Mariner 9 could not distinguish between features which had less than 5 per cent albedo contrast, that Mars was only observed for half a Martian year, or the violent dust storms that preceded Mariner 9's visit caused the real truth to be hidden.'

It was an intriguing statement – and one that was to be echoed in some discoveries on the planet to which we shall come later.

There was, however, no doubt about some of the remarkable geological features that the Mariner probe pin-pointed on the surface. Foremost among these were enormous volcanoes and immense canyons – but let the NASA report give us the facts.

'From January 1972 onwards,' says the report, 'every week was punctuated by new and startling discoveries. First there were enormous volcanoes standing as much as 15 miles above the average surface, and each one about the size of Arizona. (*Vide* Olympus Mons, with its long finger-shaped lava flows on its flanks, 15.5 miles high and 435 miles wide – the largest volcanic pile ever photographed.) Then, totally unanticipated, canyons appeared, including a great equatorial chasm more than ten times the size of the US Grand Canyon. (*Vide* Valles Marineris, 1,550 miles long, 125 miles wide and over 2 miles deep – comparable in size to the East African Rift Valley system and the same distance as from New York to Los Angeles!)

'About half the surface of the planet was shown to consist of ancient cratered terrain surrounding large impact basins. The largest circular feature, Helas Planitia, is almost twice the size of the largest basin on the moon, Mare Imbrium. The remainder of the surface is covered by younger volcanic rocks and volcanoes.'

The report continued with barely concealed delight to discuss the great lowland area, Chryse Planitia, into which broad, sinuous channels descended. 'The canyons proved to have eroded walls,' it said, 'and in addition numerous dendritic tributaries extended back from the canyon walls, suggesting that water erosion may have played a role in sculpturing the surface of Mars some time in the past. Yet it was known from previous fly-by missions that atmosphere and surface temperature conditions are such as to prevent liquid water from existing in adequate quantity at the present time. For this reason the science team was astounded by the apparent evidences of erosion, and then by the discovery of non-canyon-related sinuous channels that had all the earmarks of dry river valleys. Eroded cliffs appeared, as well as wind-erosion features and large dune masses. It is difficult to convey the sense of high excitement that pervaded the scientific investigators as the newly perceived character of our sister planet began to unfold.'

Highly emotive language from an organization usually given to the most unemotional and objective reporting! Yet it clearly revealed *why* the NASA team had to modify or abandon almost all the generalizations they had built up about the red planet. It was a bigger surprise packet than anyone had even dared imagine. And more varied and dynamic, too!

Naturally enough, these discoveries revitalized all the talk about life on Mars. For if those sinuous channels once carried water, then the planet had – might even *still* have – a basic requirement for life. An alternative explanation, of course,

96

Top Left: The great Russian enthusiast, Konstantin Tsiolkovsky, in 1910, at the time he was writing his essays about the planet.

Top right: An early photograph of Robert H. Goddard at work on his rocket experiments.

Above: The first photograph ever taken on the surface of Mars by the Viking 1 lander just minutes after it had touched down.

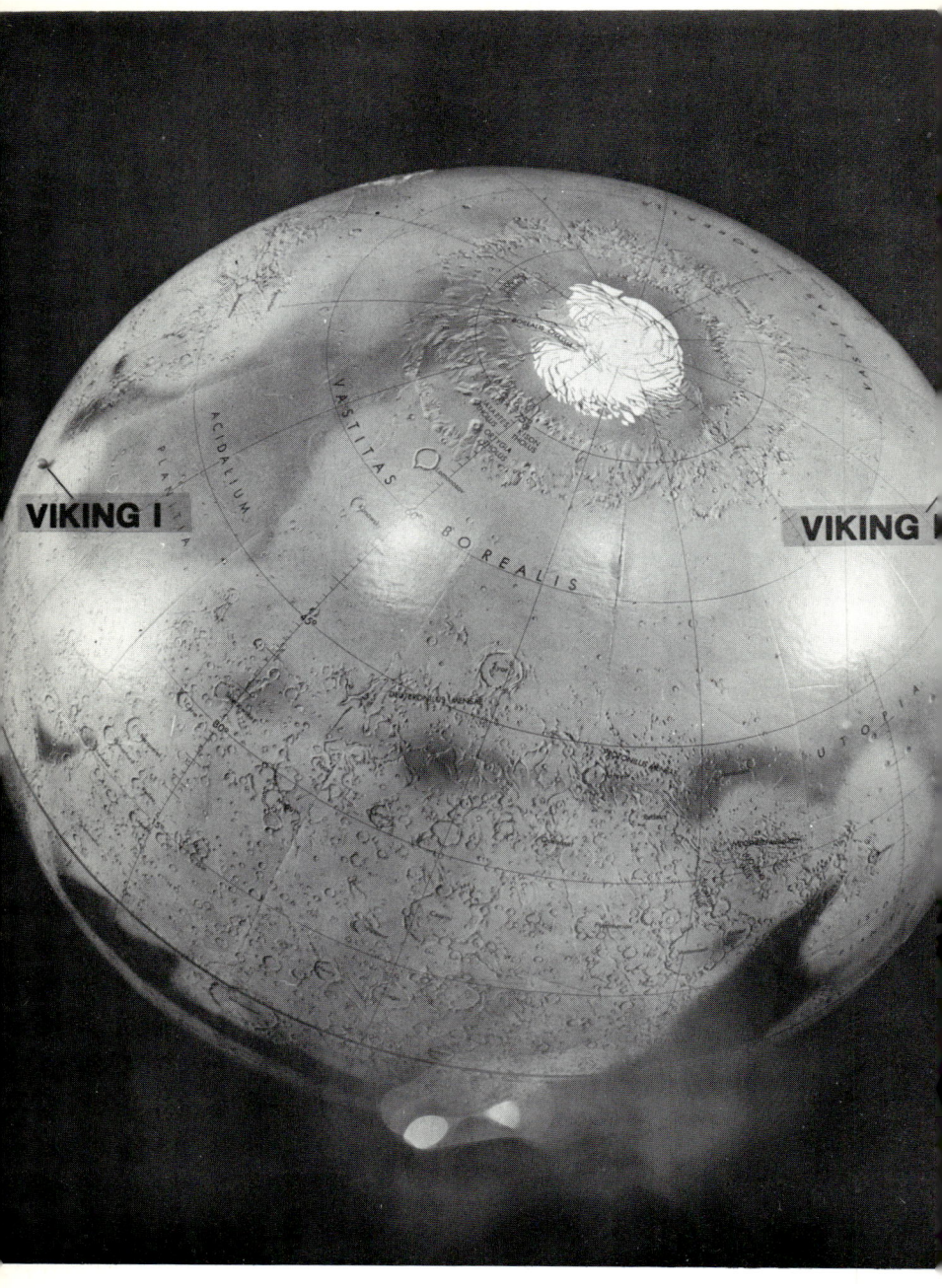

VIKING I

VIKING I

A globe of Mars prepared by NASA to show where the two Viking landers were to touch down in 1976.

Top: August 1976 and Viking Orbiter 2 approaches the dawn side of Mars. The two distinctive features are Ascreaus Mons, a giant volcano with water ice cloud plumes on the left, and Argyre, a huge crater basin.

Above: The crater named Lomonsov in the north polar area which would make a most appropriate landing point for the first Russian manned expedition to Mars.

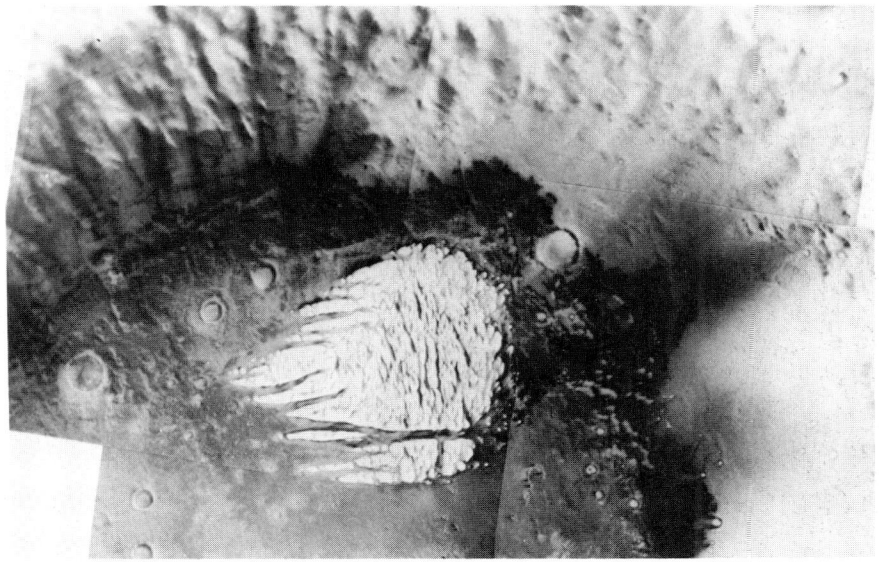

Opposite: The Viking lander, now still and silent on the surface of Mars, awaits the first human beings to step on to the planet.

Top: The smaller Martian moon, Deimos, pictured by the passing Viking spacecraft at a range of 3,300 kilometres.

Above: White Rock, one of the many strange features on Mars, photographed in the planet's equatorial regions by the Viking Orbiter in September 1978.

A stunning view of almost 12,000 miles of Martian terrain taken by the Viking Orbiter in July 1976. In the foreground is the vast plain known as Argyre Planitia surrounded by several huge craters.

Top: Man gets his first close-up look at the surface of Mars by courtesy of the cameras of the Viking lander in July 1976. This stereo view is looking south-east across the area known as Chryse Planitia.

Above: The curious letter 'B', one of several such inexplicable signs photographed on the surface of Mars by the lander.

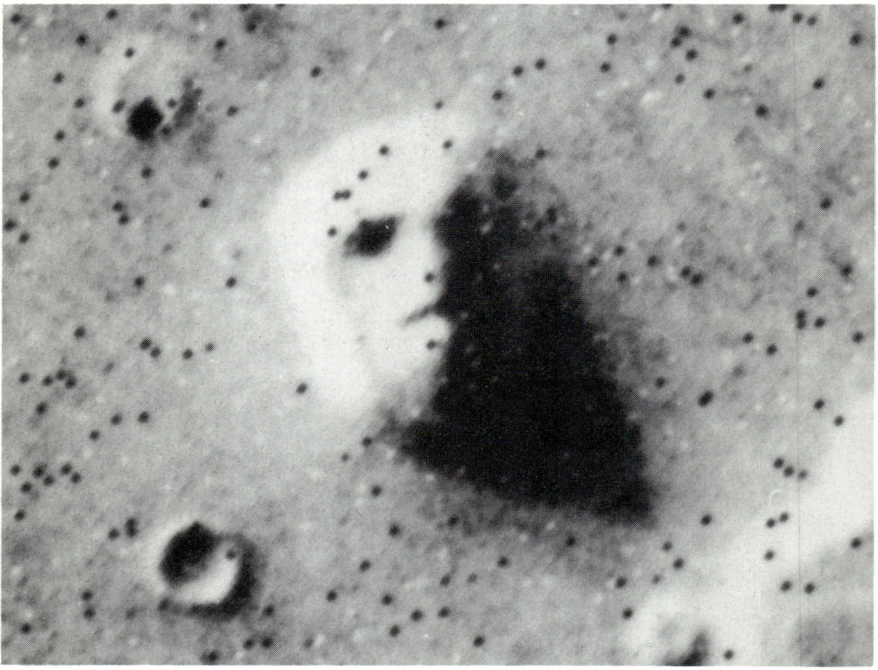

Top: The most tantalising of all photographs from Mars – the 'face', pictured by the Orbiter in the northern latitudes.

Above: An enlargement of this Martian feature which, it is being argued, may once have been part of an ancient civilisation.

was that the permafrost observed near the poles could have melted and flooded certain areas of the planet's surface – but not remained long enough to sustain the evolution of any kind of life.

Among the other Martian enigmas that the Mariner cameras pictured were a number of large circular basins such as Argyre Planitia and Isidis Planitia, which have been called the oldest recognizable structures on Mars and may well have been formed by impact during the actual creation of the planet. Photographs also revealed that the wind which had caused the initial delay to the probe's observation of Mars was hard at work 'sculpturing' the surface – moving silt and clay from one area and redepositing it in others. It was this action, said the NASA observers, which explained the changes in the surface markings of Mars that had puzzled telescopic observers for generations – and occasionally inspired them to offer the most romantic solutions.

By the time the scientists had finished sifting through the thousands of images of Mars sent back by Mariner, they couldn't help but admit that the long-held ideal that there were features on Mars very similar to those on Earth was indeed true. Could it be, they also wondered, that the Earth's environment might be destined ultimately to become like that of Mars?

Inevitably, there were as many questions left unanswered as answered, though one thing *was* certain – and Carl Sagan put that neatly into words as the 516-day mission came to a close on 27 October 1972, when Mariner 9 finally ran out of attitude control gas and was left silently circling the alien world it had so spectacularly revealed to human eyes.

'Mariner 9 has confirmed that true changes occur on Mars,' he said. 'These changes are best explained in terms of wind-blown dust, and do not require a biological explanation. Of course, this does *not* demonstrate that life does not exist on Mars; the only way to settle that argument is to land on the surface and look.'

Those words were the incentive to launch a still more ambitious project of missions to Mars – the Viking craft which in the mid-seventies would not only circle the red planet but achieve that ancient dream of successfully landing upon it.

Just how ambitious the project was to be was underlined at the outset by the Associate Administrator, Noel W. Hinners. 'Not so long ago,' he said, 'the idea of taking pictures of Mars from its surface was an idea located intermediary between far out and preposterous. It has changed from a dream to a concept with the advent of the Viking programme. As so often in the exploration of space, we were able to push back the boundaries of the practicable.'

The Viking Mission was set up by NASA in 1968, taking the place of an earlier project called Voyager which had been talked about since 1965 but was abandoned because it 'lacked both scientific merit and exploratory excitement.' Viking, in contrast, was ambitious almost to the point of audacity: two Orbiter-Landers were to be placed in orbit around the planet and having searched for suitable landing sites would then release the landers onto the surface where they would carry out a multitude of photographic, scientific and biological activities, relaying their findings to Earth. And one of the foremost tasks of the first Mars lander – NASA decided – was to 'carry out life detection experiments to answer the question about the possibility of life on the planet.'

After virtually a decade of the most intense work by a team of 10,000 people, Viking 1 was launched from the Kennedy Space Center on 20 August 1975 and arrived at Mars on 19 June 1976. For a month it carefully surveyed the hostile-looking red world, scouting a suitable touchdown point. Viking 2 was launched on 7 September 1975, and joined its twin in a higher orbit on 7 August 1976. It, too, spent a month seeking a launch site on the opposite side of the globe. The Viking orbiters also separately studied the two Martian moons, Phobos and Deimos, of which more later.

It was on 20 July 1976 that Viking 1 released its launcher and an agonizing wait began millions of miles away for the NASA scientists. Would one error in all their hard work lead to disaster, or would their painstakingly careful calculations be rewarded by a safe landing? They need not have worried.

For just seven years after astronauts had first landed on the moon, a new inhabitant arrived on Mars.

It was a milestone in space exploration – and a moment that Thomas A. Mutch, the leader of the Viking team, would never forget. 'Tim' Mutch, as he was known to his friends, later recalled the final minutes of countdown on that July day as the six-sided, work-bench-like lander spiralled downward into the late afternoon Martian sky, controlled partly by a parachute (which it ditched before touchdown) and its three retro-rocket engines. His account of how he felt watching those historic moments from the Jet Propulsion Laboratory in Pasadena, California is as precise as one would expect from such an exceptional man.

'5 a.m. The final descent begins. Conversation stops – an overwhelming silence. We listen to the mission controllers as they call out each event. After years of waiting, hoping, guessing, the end rushes towards us – too fast to reflect, too fast to understand.

'5.05 a.m. 400,000 feet.

'5.09 a.m. 74,000 feet.

'5.11 a.m. 2,600 feet.

'5.12 a.m. Touchdown. We have touchdown!

'It worked! Amazingly, it worked. Everywhere people are cheering, shaking hands, embracing. I decide not to join the celebration. It is too soon. Forty minutes more remain before the first picture from the surface of a far planet will assemble on the television screen.

'5.54 a.m. I study the blackness of the television screen, waiting for the narrow strip that will signal the first few lines of the first picture. And it appears. A sliver of electronic magic. Areas of brightness and darkness. The picture begins to fill the screen. Rocks and sand are visible and – finally, at the far right – one of the spacecraft footpads, a symbolic artifact that stamps our accomplishment with the sign of reality.

'Time and time again I repeat, "It's incredible". And truly it is. Nothing before or after can compare. It is transparent, brilliant, boundless. An explorer would understand. We have stood on the surface of Mars.'

This very first landfall on Mars had taken place at the appropriate name Plain of Gold in the midst of the smooth terrain known as the Chryse Planitia (Plains of Chryse). Once

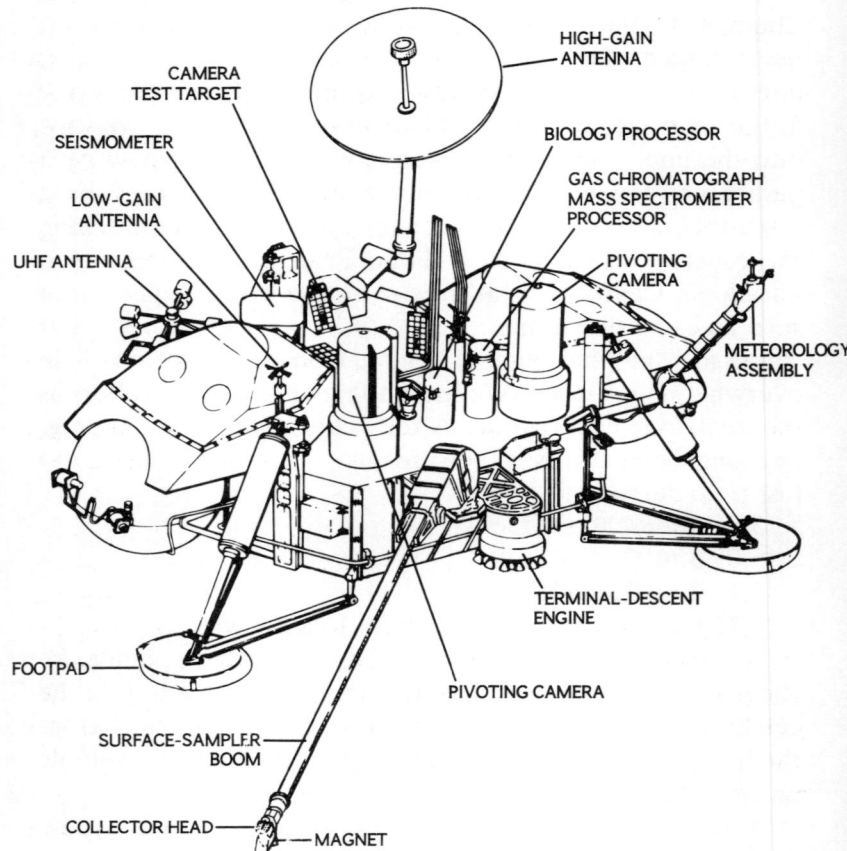

CAMERA TEST TARGET

SEISMOMETER

LOW-GAIN ANTENNA

UHF ANTENNA

HIGH-GAIN ANTENNA

BIOLOGY PROCESSOR

GAS CHROMATOGRAPH MASS SPECTROMETER PROCESSOR

PIVOTING CAMERA

METEOROLOGY ASSEMBLY

TERMINAL-DESCENT ENGINE

FOOTPAD

PIVOTING CAMERA

SURFACE-SAMPLER BOOM

COLLECTOR HEAD

MAGNET

The major component parts of the Viking lander which carried out such a multiplicity of tasks on the Martian surface.

the lander was settled, it began its mission of studying the atmosphere and geology of the planet along with analysing the soil beneath its pads. As it would never leave Mars, every discovery had to be radioed back to Earth via the orbiter circling above, and every command in return was twenty minutes in the journeying. Direct intervention in the case of any emergency was impossible; Viking was dependent on the reliability and accuracy of its programming.

NASA's own description of these two remarkable Viking landers is, I think, most informative and well worthy of inclusion at this point. 'The landers are far more complex than any spacecraft launched before,' the report says. 'Each contains the equivalent of two power stations, two computer centres, a TV studio, a weather station, an earthquake detector, two chemical laboratories (one for organic and one for inorganic analyses), three separate incubators for any Martian life, a scoop and backhoe for digging trenches and collecting soil samples, and miniature railroad cars for delivering the samples to the laboratories and incubators.

'Equipment that would normally fill several buildings had been designed in miniature to fit on a spacecraft less than ten feet across,' it continues. 'Furthermore, to avoid contaminating Mars with Earthly bacteria, the entire spacecraft was sterilized by heating it to temperatures above the boiling point of water. Each lander, and all of its one million parts, had to survive a number of major crises: the sterilization heating, the shock and vibration of launch, a one-year, 400 million mile trip through interplanetary space, the passage through Mars' atmosphere, and the landing on its surface. No wonder there were heartfelt cheers from the scientists and engineers when Viking 1's first pictures began to appear!'

No wonder, indeed. But it was what the lander produced when it went about its business of studying the planet that really captured everyone's imagination.

Of course, the safe landing immediately proved one thing about Mars. Its surface, a mixture of thickish soil scattered with rocks of all sizes, was substantial enough to support a heavy piece of equipment. The onboard cameras also revealed that the soil was reddish grey in colour, while the rocks varied in shade from dark to light grey, a few even having a red hue. They looked very like the lavas produced by erupting gas-rich volcanoes on Earth, and the scientists came to the conclusion Viking might well have landed on ancient Martian lava flows. The observers back on Earth were also quick to notice a startling similarity between the planet's gently rolling vista and a desert scene in the American southwest.

The Utopia Planita (Plains of Utopia) where the second Viking landed was, if anything, more undulating, with the rocks being still more abundant though uniform in size. This more northerly site enabled the lander to observe the enigmatic polar regions at closer quarters. Both landers charted the variable weather conditions, registering winds which could gently blow up dust as well as rippling the vast sand dunes. One of the huge storms which Mariner 9 had observed from orbit also blew across Mars in the spring of 1977 and gave both landers (as well as the scientists on Earth) an insider's view of a Martian storm.

The two machines reported that although Mars' atmosphere is thinner and colder than Earth's, as well as being persistently hazy (caused by suspended red dust which gives the sky a creamy-pinkish hue), there were, as had been expected, some striking similarities. I quote from a NASA report of 1977:

'Some of the similarities were surprising because the atmosphere of Mars is less than a hundredth as dense as Earth's. Nevertheless, on both planets, the atmospheric temperature reached its peak at about 3 p.m. local time. The daily temperature variations recorded by Viking showed the same pattern as records from a terrestrial desert "control" site at China Lake, California, although the temperature in the two places differed by more than 83 °C (150 °F). Furthermore, the changing patterns of wind direction over the flat Plains of Chryse on Mars were duplicated by the winds blowing over the equally flat Great Plains of the mid-western United States.'

The report also added that the Martian weather included two other features very familiar to us – fog and clouds. 'In small valleys,' it said, 'atmospheric water freezed out during the Martian night and then vaporises again when the sun rises, forming local patches of white fog that varnish quickly in the relative warmth of the Martian day. And though the Martian air contains only about 1/1,000th as much water as Earth's atmosphere, even this small amount can condense out, forming clouds that ride high in the atmosphere or swirl around the high slopes of Martian volcanoes.'

But the main objective of the Viking mission was, of course, to search for life forms on the planet – to determine whether the soil of Mars was dead like that of the moon or teeming with microscopic life like the rich terrain of Earth. After all, evidence had been building up with each mission. The planet might be arid, but there was permafrost at the poles, water had apparently flowed in the past, and with the discovery of nitrogen in the atmosphere, all the basic building-blocks of life – oxygen, hydrogen, carbon and nitrogen – were there.

In order to carry out these experiments, a special extending arm of the lander scraped up samples from the surrounding terrain, brought these into the lander and then divided them into three separate biological 'laboratories' for testing. It was assumed – because what else could the scientists do? – that Martian life would react like Earth life, and because no evidence of visible life had been observed, that all the effort should be concentrated on detecting micro-organisms.

In the 'laboratories' were sensitive instruments designed to detect carbon-based Martian microbes or similar creatures living in the soil. The compartments were essentially incubators, and as such designed to warm and nourish any life, living or dormant, in the samples, and to detect the chemical products of the organisms' activity.

In the first of these, designated *Carbon Assimilation*, the characteristic of terrestrial organisms such as plants to transform carbon dioxide in the air into organic compounds was simulated on the soil with radioactive CO_2. In the second, the *Gas Exchange*, nutrients and water were added to see if, like our living organisms, the Martian soil gave off gases to evidence biological activity. The *Label Release* experiment mixed a variety of radioactive nutrients to the soil to test whether this 'food' would be consumed to give off carbon dioxide as terrestrial animals (including humans) do when consuming organic compounds.

It was an agonizing wait for some kind of result from the tests – and not a completely satisfying one when it arrived. But neither was it without optimism. Let Tim Mutch put it in his own inimitable way.

'Since all three of the experiments designed to test metabolic activity of a microbiota yielded "positive" results, it is tempting to conclude that life exists on Mars,' he said in April 1978. 'Indeed, some of the early meetings, in which preliminary biology results were reported, were charged with the excitement of profoundly important speculations about to become historical reality.

'We now recognize that the biological results can be explained by inorganic surface reactions in the absence of any living forms. Strengthening this conclusion is the absence of organic compounds, documented by the gas experiment.

'But,' he added, 'surely this does not prove the absence of life on Mars, only its absence at two localities purposely chosen to be bland and featureless. It remains possible – perhaps unlikely, although statistics in this instance have little validity – that life exists elsewhere on Mars in some special environmental niche – or that it existed millions of years ago.'

I shall be pursuing this fascinating speculation again later. (Incidentally, before going any further I should just like to record that the achievements of the late Tim Mutch are to be recognized in a unique way when man finally does reach Mars. For NASA has decided that *when* the first astronauts land, the task of one man will be to visit the now still and long-silent Viking 1 on Chryse Planita and affix a plaque to it, renaming the lander 'The Thomas A. Mutch Memorial Station' in honour of the man who had done so much to enhance our view and knowledge of Mars.)

While, however, all this activity was going on down on the surface of the red planet, the two orbiters circling above were also adding vastly to the store of information and photographs about the planet – as well as providing the first-ever detailed look at Mars' two extraordinary moons.

Even before the landers had touched down, high resolution photographs were on their way to earth of the planet's major features – the huge volcanoes, the great canyons and landslides, the mountains and circular basins, the craters and vast, winding water-cut channels. Then, in September 1976, the orbit of Viking 2 was changed so that it could circle over the mysterious white polar regions of Mars and confirm the scientists' growing belief that these caps consisted of water ice instead of CO_2 or 'dry ice'. This discovery – plus the fact that steep-sided valleys were spotted in the ice – also added weight to the conviction that melted water from here may well have cut the channels on Mars in times long gone by.

The final sequence of the mission took place in February 1977 when the orbits of the two Vikings circling the planet were altered to allow them to make close approaches of under 100 miles each to Mars' two moons, Phobos and Deimos. It was a survey carried out with a nice acknowledgement to history, for it was just one hundred years earlier, of course, that the two little satellites had first been discovered by the American astronomer, Asaph Hall. One couldn't help wondering, as the two orbiters homed in on the oddly-shaped moons, whether Hall had ever dreamed that only a century after he had spotted them indistinctly through a telescope they would be being studied close-up by a man-made machine?

In fact, the high resolution images that the orbiters sent back to Earth were as good as any obtained during the fly-by or orbiting of any body in our solar system. The pictures also changed a number of our ideas about the satellites, both of which are locked into a stable, synchronous orbit of the planet around its equatorial regions at 3,700 miles (Phobos) and 14,577 miles (Deimos) respectively.

Phobos, with its irregular shape of about 8.4 miles in width and 5.96 miles in height, was found to be rather smaller than expected, while asteroid-shaped Deimos at 4.66 miles long, 3.73 miles wide and 3.4 miles high was somewhat larger. The Viking cameras revealed that both were uniformly grey in colour, and with albedos of .06 are among the darkest objects in the solar system: hence the time and difficulty it took the old astronomers in finding them. Both are heavily cratered, although Deimos has the smoother appearance of the two.

The eyewitness observation of Phobos was particularly instructive, because it is one of only three satellites in our solar system whose rotation period of 7 hours 39 minutes is less than that of the primary planet – Mars' period, of course, being almost exactly identical to the Earth. This means the moon travels around its mother *twice* in a Martian day! Viking was also able to confirm for the scientists that Phobos is actually losing orbital energy to the surface tides on Mars. And as this orbit decays and it gets closer to Mars, there is every likelihood that it will be torn apart when the tidal forces of Mars overcome the cohesive bond between its particles. As it is already inside the 'Roche Limit' where internal gravity alone is too weak to hold it together, the experts believe it could disintegrate into small particles and form a 'ring plane' around Mars within the next 50 million years. (Such as those discovered around Jupiter and Uranus.) Either that, or it could fall into the surface of the red planet itself.

At the present time, though, the surface of Phobos is dominated by sharp, fresh-looking craters of all sizes and a vast network of linear features resembling crater chains. These massive grooves several miles long appear to be associated with the planet's largest crater, named Stickney, which is about ten kilometres in diameter. Similar networks were observed by Viking on Deimos, though they and the craters seemed to be partly obscured by debris. A large depression near the South Pole is believed by scientists to have been caused by an impact.

The debris material found on both moons, and known as regolith, has particularly interested the scientists. For informed speculation has it that oxygen and other compounds could be synthesized from this material and then turned into fuel for use in manned expeditions on either Phobos or Deimos – even Mars itself. For this reason, mineral exploration on the satellites has been earmarked as an important consideration in any future expedition to this part of the solar system.

Just as intriguing is the puzzle as to how Phobos and Deimos originated. Put simply, they could either be asteroids 'captured' by Mars at some time long ago in its history, or else they are made from the debris left over from the actual formation of the planet. With the Viking missions now over, only further survey by a lander spacecraft can really hope to solve this mystery.

Such an investigation is also required on the red planet

itself to examine undoubtedly the strangest of all the surface features that the Viking probes have brought to light: features that some experts have claimed indicated Mars was *once inhabited*. They provide the focus of my next chapter – and might well be called the most compelling reason of all for mankind to reach Mars. . . .

SIX

The Search for Martians

THE POSSIBILITY that life of some kind might exist on Mars has, of course, been of widespread public interest for years, encouraged in our times by the reports of Giovanni Schiaparelli and Percival Lowell, and then raised to fever pitch by the publication of *War of the Worlds*. One of our leading scientific authors, Isaac Asimov, wrote recently:

'A man most influential in convincing the public that there *was* intelligent life on Mars was the English writer H. G. Wells who responded to the growing popularity of the Schiaparelli/Lowell view by publishing *The War of the Worlds* in 1898. In this book Wells told of Martians who, despairing of being able to maintain their dying world, launched an invasion of Earth. Their superior technology made it possible for them to overwhelm the inhabitants of Earth, but the Martians were defeated by physiology, for their bodies could not resist the onslaught of Earth's decay bacteria.'

Though today we can be sure that there are no even remotely humanoid inhabitants of the planet, the possibility that there *might* once have been is being seriously advanced as a result of some extraordinary photographs taken during the Viking mission. These we shall consider after a look at the history of man's fascination with the idea of Martians.

Although records are unclear as to just when it was postulated that Mars might be inhabited, the first serious attempt by mankind to make contact with supposed dwellers on the red planet can be traced back as far as 1820. The man to propose this scheme was no crank, but a distinguished and influential German astronomer and mathematician, Johann Karl Gauss (1777–1855), who had devoted his life to the study of the solar system and actually invented new methods for the calculation of the orbits of the planet.

Gauss' fascination with Mars – in particular what he saw through the swirling clouds which mostly obscured, but occasionally revealed the planet's surface – led him to the conclusion that if there were beings observing Earth as he watched them, then some vast sign or emblem on our world might be the first step towards communication. And so he proposed the growing of a huge triangle of wheat surrounded by pine trees on the wide-opened spaces of Siberia. Though this idea was apparently welcomed in some German scientific circles, all attempts to enlist the aid of the Russians in the scheme came to nothing.

A perhaps even more grandiose scheme was put forward thirty years later by a French astronomer, Charles Cros (1801–83), who elaborated on Gauss' idea of signs by suggesting building a huge mirror to focus beams of sunlight onto Mars. Not content with using this to effect a kind of interplanetary Morse code, Cros wanted the mirror actually to burn characters and numbers into the Martian sand! His scant concern for any unfortunate inhabitant who might be scorched to a cinder by such a scheme did not endear him to the more humanitarian of French scientists!

A more practical idea was advanced by Nikola Tesla (1857–1943), the Yugoslav-born American inventor who, after studying at Graz, Prague and Paris emigrated to the USA in 1805. After some years at the Edison Works at Menlo Park, he quit to develop his own inventions which were to include improved dynamos, transformers and electric light bulbs as well as the high-frequency coil which bears his name.

In 1899, with the Mars fever just beginning to develop after the appearance of *War of the Worlds*, Tesla came up with an idea he hoped would instantly resolve the issue of life on the planet. Aided by a giant version of the electrical coil he had invented, he proposed to try and 'talk' to the red planet. It was a bold scheme, and one that attracted considerable public attention, for Tesla's laboratory at Pike's Peak in Colorado was becoming widely regarded as the place where the impossible became possible.

In essence, Tesla proposed to generate powerful electromagnetic surges from an electrical coil 70 feet in diameter. These he would utilize to contact the far-distant planet in a way rather like the telegraph. Although the American scientific establishment was far from convinced of the practicality of the idea, large numbers of ordinary people urged Tesla at least to have a try. Unhappily, though, when the system was complete and tried out it only succeeded in making all the light bulbs glow in a 25-mile radius of Pike's Peak!

These failures did nothing to dampen public enthusiasm, however. Hardly a year passed without an astronomer somewhere in the world reporting strange phenomena he had observed on Mars. Ephemeral lights and patterns were regularly reported, as were light and dark spots, and most startling of all – brilliant flashes of light. These last intrigued the observers most, for they were evidently of huge size, otherwise they could not have been seen across 35 million miles of space. What could these giant electrical phenomena be, laymen and scientists alike wondered? The evidence of advanced technology, suggested imaginative writers – or signs of active volcanoes, wondered the more cautious scientific establishment?

A report in the American scientific journal, *Observatory*, of September 1894, is typical of the period. 'A telegram from M. Perrotin of Nice on August 6 announcing the observation of a bright prominence beyond the terminator of Mars, and a subsequent announcement by Mr Stanley Williams in America of similar observation, have renewed scientific interest in this class of phenomenon, and have also created some sensational interest in the public press. M. Perrotin says that these were seen with great distinctness, and that it is scarcely possible to consider them the results of illusion.'

The journal continues, 'The first hint of such bright markings came from Professor Schiaparelli who, in 1888, in publishing some observations of white spots stated that the whiteness was always more pronounced when the spots were near the edge of the disk, but he did not observe the brightness beyond the terminator. In the same year, M. Terby observed similar white spots, which were invisible until they approached the western edge of the disk, when they appeared very bright, and were apparently seen beyond the edge of the disk by irradiation; but it was not until July 1890 that the phenomenon of obvious projections was seen. It is to be remarked that the spot seen on July 6 was situated at the end of a long, bright stripe on the surface of the planet.'

And turning to the most recent sighting, *Observatory* added, 'As to the cause of these, there appear to be two rival explanations. One, as suggested in Mr W. H. Pickering's report of his observation, that these luminosities are bright clouds; and in favour of this theory he says that the small mass of the planet is not inconsistent with clouds at such great altitudes as these observations would require. On the other hand, the observations show in many cases that these bright objects are permanent on the planet, a quality which we find difficult to attach to clouds.'

Such lights paled into insignificance against those reported in 1900 which fanned the Mars fever into still greater prominence when the suggestion was made that the Martians were now trying to get in touch with *us*!

On December 27, the French astronomer, M. de Fonvielle, reported to the daily newspaper, *Le Matin*, that he had observed the previous evening 'a series of bright lights which suddenly appeared on the surface of Mars in a straight line extending for several hundred kilometres. These gigantic fires,' as he described them, 'burned without interruption for one hour and ten minutes, and then disappeared as suddenly as they had come.'

The astronomer's conviction that these fires had been lit by Martians delighted headline writers around the world, but did not impress the British *Astronomical Journal* which declared:

'M. de Fonvielle evidently believes in a message from Mars, and says that if the Martians really lighted these fires, it is indispensable for the astronomers of this world to let them know that they have been understood, and that we count on their intelligence to succeed in understanding us and creating an alphabet.'

The journal dismissed the reported fires as being 'probably caused by brilliant clouds projecting beyond the terminator or illuminated tops of ranges of mountains on the planet.'

In direct contrast to such scepticism, in the following year, 1901, a prize of 100,000 francs was offered by a certain Madame Clara Goguet, the wife of a wealthy Paris industrialist, for the first person to make contact with extra-terrestrials from anywhere *but* Mars! Apparently, the good lady and her supporters were convinced that contact with Martians was imminent and therefore too easy to qualify for the award!

But the de Fonvielle sighting was to prove not the only alleged message from the red planet. Two years later, in April 1902, what looked very like a sequence of light and dark flashes was observed by an American astronomer, James Campbell, who described them as 'demonstrating some of the qualities of an Indian smoke signal.'

Initially, the world's press had the good sense to check and make sure this was not an April Fool's joke, but Campbell stuck to his guns about what he had seen. A solution was not, however, long in coming – and from the man who most strongly promoted the idea of life on Mars, Percival Lowell. In a statement issued from his Flagstaff observatory he said:

'The origin of this phenomenon appears to have been in the locality of the area known as Icarium Mare. Now the form of the cloud was of the same general shape – a broken cloud stretching east and west five times as far as it did north and south.

'The Icarium Mare is undoubtedly a great tract of vegetation,' he continued emphatically, 'where moisture would be held and whence it could accordingly be given off. Arising there, either from seasonal or temporal cause, the vapour would gather into a cloud or clouds and proceed to float away over the desert regions of the north. If this, then, is what happened, we may conceive the cloud as having been generated over the Icarium Mare, rising to a height of thirteen miles, and then travelling east by north at about twenty-seven miles an hour into the desert of Aeria, there to dissipate after an existence of three or four days. That it was a phenomenon of capricious not regular production is shown by its not having been repeated – that is, it partook of the subtle unpredictability of cloud.'

Though we now know, of course, that there is no vegetation on Mars, it is very much to Lowell's credit that despite his obsession with the idea of life on the planet, he did not clutch at straws like this to support his case.

One of the most famous personalities to become involved in this period of Mars history was the great Guglielmo Marconi, the 'Father of Wireless'. Marconi (1874–1937), the Italian-born inventor, began successful experiments with wireless telegraphy in Italy and then came to England where he first succeeded in sending signals across the Atlantic in 1901. It was in 1921, however, that Marconi astonished the world by announcing that he believed he had intercepted some messages from the red planet.

News of this incident was first revealed in September 1921, when Mr J. C. H. MacBeth, the London manager of the Marconi Wireless Telegraph Company, arrived in New York and told reporters of the mysterious signals from outer space that his boss had picked up on his equipment.

The signals, MacBeth said, had been received while Marconi was on his yacht in the Mediterranean Sea conducting atmospheric experiments with wireless. Some magnetic wave-lengths high in the metre band had been picked up, although the maximum length of Earth-produced waves at that time was 14,000 metres. Any idea that the waves might have been produced by electrical disturbances was disproved by the regularity of the impulses, said MacBeth.

And although the impulses apparently consisted of a code, he added, the only signal similar to Earth codes was one resembling the letter V in the Marconi code. Were these, then, messages from Mars? It was a solution the believers liked, but not one to satisfy the scientists.

With the further development of radio, a number of interesting discoveries were made which seemed to offer a more likely means of contacting Mars than any of the earlier rather eccentric proposals. At least that was what the newspapers and magazines of the nineteen twenties repeatedly told their readers.

In 1922, Mr L. W. Chubb, the director of research for the American Westinghouse Electric Company, announced the successful development of beam radio transmission, and said that if communication with Mars was ever established 'it would have to be with ultra-short waves directed like a beam of light in order to penetrate the atmospheric layers above the Earth's surface.'

Ultra-short waves, he explained, were the nearest approach of radio waves to regular light waves, and were the only ones capable of breaking through the layers of ionized gas around the planet which reflected all normal radio medium and short waves.

Although Westinghouse was not prepared to carry out further experiments, a trio of European scientists did conduct some beam transmission tests with this in mind. The men, Hals, a Danish expert, and two Scandinavian scientists, Stormer and Peterson, found that certain short waves not only penetrated the ion layers but also travelled far out into space. Their signal echoes arrived from three to thirty seconds after transmission.

Since the velocity of radio waves is the same as light – 186,000 miles per second – it was calculated that the layers or bodies that reflected these signals were located at about 280,000 miles to 2,800,000 miles from the Earth. And, the team discovered, even these layers far out into space could be penetrated by a beamed wave approaching a regular light wave which passes through all ionized barriers.

It was not long after these experiments that one of the strangest, and most inexplicable, 'Contact Mars' events occurred. It is a mystery, in fact, that has persisted to this day.

The events took place during the evening of 22 August 1924, a crucial night in Mars-Earth history for the two planets were just 34,500,000 miles apart – the nearest, in fact, they will be until the year 2000. Because of this proximity, astronomers throughout the world were training their telescopes on the planet, and all those who had for years been hoping for a message from the Martians had their radio sets turned on. This was because a large number of broadcasting stations had agreed to silence their programmes periodically in the hope of picking up *something*.

Plans for this remarkable experiment had been made with

the most painstaking care. Dr David Todd, astronomy professor at Amherst College in Massachusetts and a long-time
Mars enthusiast, had organized the international 'listening-
in' test. Aided by the US government he had requested that
all countries with high-power transmitters silence their
stations for five minutes every hour from 11.50 p.m. on
21 August to 11.50 p.m. on 23 August.*

Also recruited for the experiment was Francis Jenkins, the
Washington inventor who had recently devised a radio photo
message continuous transmission machine. He was asked by
Dr Todd to make a record of any signals received during the
twenty-four hours.

The Jenkins recording device was attached to a receiving
set adjusted to a wavelength of 6,000 metres. Incoming
signals caused flashes of light which were then printed on a
film in the form of a roll tape, thirty feet long and six inches
wide.

All was in readiness as Mars neared its closest point to the
Earth. The silence was almost uncanny as men and women,
laymen and scientists alike, all over the world held their
breath.

Then, suddenly, in the midst of this almost ethereal silence,
came some eerie signals, the like of which had never been
heard before. Radio Station WOR in Newark, New Jersey,
was the first to register these sounds. Almost immediately,
others picked them up, too.

In Washington, the Jenkins machine began to quietly make
a record of what was happening. When the film was
developed the following day, Dr Todd and his colleagues
knew they had not been mistaken. Something truly
astonishing had taken place.

First, there was the evidence of the listeners. They had
unmistakably heard a series of long and short beeps. But just

*Another earlier story about Dr Todd recounts how in May 1909 he was
offered the use of a balloon by the New England Aero Club in order to
ascend into the heavens and there, armed with huge antennae, listen for
wireless signals from the red planet. However, despite several trips aloft, he
heard nothing whatsoever.

as unmistakably they were not the Morse code or any other kind of code known to man.

Next, the film. This disclosed in black on white a fairly regular arrangement of dots and dashes along one side. But on the other, at almost evenly spaced intervals, were curiously jumbled groups, each taking the form of a crudely drawn human face!

Francis Jenkins was at a complete loss to explain these extraordinary features. 'The film shows a repetition at intervals of about half an hour of what appears to be a man's face,' he disclosed at the press conference called the following day to discuss the experiment. 'It's a freak which we can't explain.'

Had Mars been the cause, was the question on every reporter's lips? Had someone or *something* from the planet been trying to get in touch?

'We just don't know,' Dr Todd replied. 'But we now have a permanent record which can be studied.'

And studied it was – by Army code experts who worked on the film for some weeks. But at the end of this time, they, too, had to confess they were baffled.

In desperation, Dr Todd and his team turned to a recording of the sounds that an engineer named R. I. Potelle at Station WOR in Newark had made. Over and over again they repeated the strange signals.

Finally, after many hours of study, engineer Potelle and the Todd team concluded that it was a single word which was being transmitted. The word was 'Eunza'.

It was a word that had no meaning in any of the languages of Earth. And it is a word that has remained to this day the only real clue we have about this strange, unresolved enigma.

The failure of this experiment was not the last attempt to make contact with Mars, however. In October 1928, a London lawyer and enthusiastic amateur planet watcher named Mansfield Robinson had another go. Once again, he recruited the assistance of a broadcasting station to aid him in his attempt.

Robinson put his message out from the Rugby wireless station. It was sent on a 18,700 metre wavelength and it was his earnest hope that the signal might prompt some sort of etheric response.

Listening in while the transmission went out were a number of England's leading scientists, including Professor A. M. Low (1888–1956), the inventor, writer and president of the British Interplanetary Society.

Low admitted afterwards that he was sceptical of the whole idea, but just minutes after Mansfield Robinson's message was transmitted into space, something began to crackle through on the Rugby radio receiver.

'It was a mysterious message,' Professor Low said later. 'I thought it hardly likely that it could have come from Mars. However, I must confess that I do not know who sent it. It was a series of dots and dashes.'

Unconvinced and puzzled though the professor was, he was so fascinated by the whole experience that he afterwards used it as the basis of a science fiction novel, *Mars Breaks Through* (1937).

Two years after the Robinson experiment, yet another 'Contact Mars' project was put forward. This was a plan to establish a regular light beam signal to the planet, and was drawn up by the enigmatic English psychical researcher, Harry Price (1881–1948) whose enquiries into alleged hauntings (such as that at Borley Rectory on the Essex-Suffolk border) and investigations into witchcraft, magic and levitation had made him a famous, if somewhat showman-like, figure in the thirties. As director of the National Laboratory of Psychical Research in London, he announced in the spring of 1930 that he was hoping to open up a regular channel of communication with Mars through a colossal light signal.

The site selected by Price for the experiment was the summit of the mighty Jungfraujoch, 11,00 feet above sea level in the Bernese Oberland. There ten tons of magnesium were to be ignited in oxygen in the focus of reflectors, and the beam directed on what were believed to be snowfields around the Martian poles. This gigantic flare, Price believed, would surely bring a response of some kind if there were intelligent beings on the red planet.

Once again, though, a widely publicized and eagerly

awaited project foundered before it could be completed. The ambitious Harry Price just could not raise sufficient funds.

After so much frustrated endeavour, it might have been expected that attempts at making contact with Mars would have come to an end, and ideas of life on Mars also be relegated to the world of romance and dreams. So, indeed, it might have been but for the extraordinary events of 30 October 1938 – the night of Halloween – when a radio broadcast convinced a large section of the American public that Martians were invading their country!

As I shall explain, it was H. G. Wells' remarkable novel, *War of the Worlds*, exerting its astonishing influence on the hearts and mind of people once again!

□

The story of the 'Panic Broadcast', as it has become known, is something of a legend in radio and science fiction circles, but is well worth the retelling here because of what it reveals about the public fascination with Mars. The man behind the production was a young actor named Orson Welles who created such a realistic version of his English namesake's novel that it literally caused people to flee from their homes in panic – convinced that Martians were on the rampage!

As we have seen, Mars fever gripped America for thirty-odd years, and the recent use of radio in attempts to contact the planet had made it a particularly appropriate medium for recounting Wells' story. The fact that Orson Welles deliberately relocated the action from England to New Jersey (where, you'll also recall, the radio engineer had recorded that extraordinary message from space) and told the events in the form of news bulletins which mounted in hysteria as the threat increased, made for an absolutely convincing broadcast. Far more convincing, in fact, than Welles had ever imagined.

Like so many other people in this book, Orson Welles, who was born in 1915, had read *War of the Worlds* as a youngster, and it remained indelibly fixed in his consciousness. As early as his teens, this ebullient and self-confident young man had decided on a career in acting and by his early twenties was a seasoned stage and radio performer. Such, indeed, was his talent that at the tender age of 23, CBS Radio of New York appointed him producer of their prestigious weekly hour-long show, Mercury Theater, which specialized in dramatizations of outstanding novels. Because of his admiration for *War of the Worlds*, it was no surprise that Welles should have chosen it as ideal material. It was the mischievous streak in him that decided on broadcasting the adaptation on the night of Halloween, Sunday 30 October 1938, as Howard Koch, the man who wrote the script, later explained.

'In a sense I myself was one of the victims of the "Halloween Prank" as Orson later called it in a masterly understatement,' said Koch. 'For after listening to the broadcast in my apartment, I went to sleep blissfully unaware of what was happening outside. The next morning when I walked down 72nd Street on my way to the barber there was an air of excitement among the passers-by. Catching ominous snatches of conversation with words like "invasion" and "panic", I jumped to the conclusion that Hitler had invaded some new territory and that the war we all dreaded had finally broken out.

'When I anxiously questioned the barber, he broke into a broad grin "Haven't you heard?" he said, and he held up the front page of a morning newspaper with the headline NATION IN PANIC FROM MARTIAN BROADCAST. This was a moment that still seems unreal to me. I stared at the paper while the confused barber stared at me. Centre page was a picture of Orson, his arms outstretched in a gesture of helpless innocence, and underneath was the opening scene of my play.'

If Howard Koch felt confused then, it was nothing to what some 1,200,000 of the estimated 6,000,000 listeners to the programme had felt earlier. For according to Professor Hadley Cantril of Princeton who later made an exhaustive study of the affair, at least that number took the broadcast literally and 'reacted according to their natures and circumstances.' In addition, said the professor, an unknown number – probably several million more – who were not tuned in to the programme were also caught up in the mass hysteria.

It is, I think, worth recalling how the broadcast began. Not

119

just because of the impact it was to have, but also because of its manner of reporting Martian phenomena – a manner which was already familiar to most American listeners and certainly heightened the authenticity.

'Ladies and gentlemen,' the sonorous tones of Orson Welles as the narrator began, 'we interrupt our programme of dance music to bring you a special bulletin from the Intercontinental Radio News. At twenty minutes before eight, central time, Professor Farrell of the Mount Jennings Observatory, Chicago, Illinois, reports observing several explosions of incandescent gas, occurring at regular intervals on the planet Mars.'

And then after a pause he went on, 'The spectroscope indicates the gas to be hydrogen and moving towards the Earth with enormous velocity. Professor Pierson of the observatory at Princeton confirms Farrell's observation, and describes the phenomenon as (quote) like a jet of blue flame shot from a gun (unquote).'

This hark-back to the kind of reports of Martian phenomena we have already discussed in this chapter immediately gripped the listening audience.

The broadcast next included an 'interview' with Professor Pierson in which he was asked about the disturbances on Mars and if they were an indication of life on the planet. 'I should say,' he replied, 'that the chances against it are a thousand to one.'

When, however, a series of 'flaming objects' were reported to have fallen on the little hamlet of Grovers Mill in New Jersey, and Professor Pierson is taken to view the site by a radio reporter, he is far less sure of himself. 'I don't know what to think,' he says, examining a cylindrical-shaped object buried in the ground.

From this cylinder, a moment or two later, emerges 'something like a grey snake with tentacles' according to the horrified tones of the radio reporter. There is a sudden jet of flame and several of the assembled crowd are reduced to grisly corpses.

Thoroughly engrossed by this stage, the listening millions of Americans then heard Welles once again. Now he spoke the lines which began the panic.

'Ladies and gentlemen,' he intoned, 'I have a grave announcement to make. Incredible as it may seem, both the observations of science and the evidence of our eyes leads to the inescapable assumption that those strange beings who landed in the Jersey farmlands tonight are the vanguard of an invading arm from the planet Mars. . . . The monsters are now in control of the middle section of New Jersey and have effectively cut the state through its centre. Communication lines are down from Pennsylvania to the Atlantic Ocean. Railroad tracks are torn and services from New York to Philadelphia discontinued except routing some of the trains through Allentown and Phoenixville. Highways to the north, south and west are clogged with frantic human traffic. . . .'

And so, in reality, they were a few minutes later.

The newspapers of America bear eloquent witness to what happened in column after column of reports. Take the *New York Times* which was at the very heart of the panic. In its issue of Monday 31 October under a headline: RADIO LISTENERS IN PANIC – Many Flee Homes to Escape 'Gas Raid From Mars', it reported:

'A wave of mass hysteria seized radio listeners throughout the nation between 8.15 and 9.30 last night when a broadcast of H. G. Wells' fantasy *War of the Worlds* led thousands to believe that an interplanetary conflict had started with invading Martians spreading wide death and destruction in New Jersey and New York.

'The broadcast, which disrupted households, interrupted religious services, created traffic jams and clogged communication systems, was made by Orson Welles, who as the radio character "The Shadow" used to give "the creeps" to countless child listeners. This time at least a score of adults required medical treatment for shock and hysteria.

'Throughout New York, families left their homes, some to flee to nearby parks. Thousands of persons called the police, newspapers and radio stations here and in other cities of the United States and Canada seeking advice on protective measures against the "Gas raids".

'The switchboard of the *New York Times* was overwhelmed

by the calls. A total of 875 were received. One man who called from Dayton, Ohio, asked, "What time will it be the end of the World?"'

The files of other papers from across the nation tell the same story.

Providence, R.I. Journal: 'Weeping and hysterical women swamped the switchboard of the *Journal* for details of the "massacre". The electric company received scores of calls urging it to turn off all lights so that the city would be safe from "the enemy".'

The Boston Globe: 'One woman declared she could "see the fire" and told the *Globe* she and many others were "getting out of here".'

The Pittsburgh Chronicle: 'A man returned home in the midst of the broadcast and found his wife, a bottle of poison in her hand, screaming, "I'd rather die this way than like that!"'

The San Francisco News: 'An offer to volunteer in stopping an invasion from Mars came among hundreds of telephone enquiries to police and newspapers here during the radio dramatization of H. G. Wells' story. One excited man called Oakland police and shouted, "My God! Where can I volunteer my services? We've got to stop this awful thing!"'

Indianapolis Tribune: 'A woman ran into a church screaming, "New York destroyed – it's the end of the World! You might as well go home to die. I just heard it on the radio." Services were dismissed immediately.'

Atlanta Journal: 'Listeners throughout the South called newspapers reporting that Martians had struck New Jersey and anywhere from 40 to 7,000 people had been killed. Editors said that responsible persons known to them were among the anxious information seekers.'

There were even stories of unintentional humour.

The Reno Telegraph: 'Marion Thorgaard, here for a divorce from Robert Thorgaard of New York, collapsed, fearing her mother and children in New York had been killed. One man immediately started East in the hope of aiding the wife he was here to divorce.'

Los Angeles Times: 'One woman told a reporter of this newspaper that her only thought was delight that the Martians were coming, "I won't have to pay the butcher's bill," she said.'

Without any doubt, the broadcast had been more realistic than anyone could have imagined – and Orson Welles found himself the centre of a storm of protest. Though it made him famous overnight, he preferred to refrain from comment, leaving that to his assistant, John Houseman, who said a few days later:

'I remember during the playing of the final theme, the phone started to ring in the control room and a shrill voice through the receiver announcing itself as the mayor of some big Midwestern city. He is screaming for Welles. Choking with fury, he reports mobs in the streets of the city, women and children huddled in churches, violence and looting. . . .

'Orson hangs up quickly. For we are now off the air and the studio door burst open. The following hours are a nightmare. The building is suddenly full of people and dark blue uniforms. We are hurried out of the studio, downstairs, into a back office. Here we sit, incommunicado, while network employees are busily collecting, destroying and locking up all scripts and records of the broadcast.

'Hours later,' Mr Houseman adds, 'instead of arresting us, they let us out a back way. We scurry down to the theatre like hunted animals in their holes. It is surprising to see life going on as usual in the midnight streets.'

Despite the undoubted upset the broadcast caused, it has now become accepted that it actually did more good than harm. For in those uneasy days of 1938 when war was, of course, looming on the horizon, it clearly demonstrated how vulnerable people were to a panic reaction in the event of war. The nationally-syndicated columnist, Dorothy Thompson, pin-pointed this precisely in an article shortly afterwards.

'All unwittingly,' she said, 'Mr Orson Welles and the Mercury Theater have made one of the most fascinating and important demonstrations of all time. They have proved that a few effective voices, accompanied by sound effects, can so convince masses of people of a totally unreasonable, completely fantastic proposition as to create nationwide panic. They have demonstrated, more potently than any argument, demonstrated beyond question of doubt, the appalling dangers and enormous effectiveness of popular and theatrical demagoguery.

'And far from blaming Mr Orson Welles, he ought to be given a Congressional medal and a national prize for having made the most amazing and important contributions to the social sciences. For Mr Welles and his theatre have made a greater contribution to an understanding of Hitlerism, Mussolinism, Stalinism, anti-Semitism and all the other terrorisms of our times than all the words that have been written by reasonable men. They have made the *reducto ad absurdum* of mass manias.'

But if the 'Panic Broadcast' left a nation somewhat red-faced, it did nothing to halt the speculation about life on Mars – only brought it to a temporary halt during the grim years of the Second World War.

It is, incidentally, just worth noting that the *War of the Worlds* has since been re-broadcast no less than three times with not dissimilar effects. In 1955 an adaptation by Russian radio so startled the population that a Soviet astronomer had to hastily call a press conference to deny that Martians had invaded the country. In 1971, another American radio station in Buffalo put out a revised version of Welles' broadcast, also on Halloween, but re-siting the story in its own locality. Despite the fact the programme had been promoted weeks in advance as a fake, the police switchboard 'immediately lit up like a Christmas tree with callers' according to one report. And in 1984 in London when Capitol broadcast the original 1938 programme, they, too, were deluged with anxious calls – including, it is said, one from the American Embassy. 'Will they never learn?' one exasperated journalist asked the following day.

It was not long after the war that interest in life on Mars was revived. The main cause this time was the alleged sighting of what were called Unidentified Flying Objects, or more popularly 'Flying Saucers'. According to a number of the stories which subsequently emerged these strange machines came from space – not a few of them from Mars and piloted by 'little green men'.

The whole saga of the UFOs began in June 1947 when a young American pilot named Kenneth Arnold was flying his small plane near Mount Rainier in Washington. Suddenly he saw above him a group of flat, disk-shaped objects which he later described as flying 'like a saucer would if you skipped it across the water.' In recounting his strange sighting, the newspapers coined the term 'Flying Saucers' which has continued to be used for such craft ever since. And the number of alleged sightings as well as encounters with these so-called spaceships and their occupants now fill a small library.

Right from the onset of the phenomena, Mars was placed high on the list of planets from which the UFOs might originate – particularly as some researchers went to great lengths to try and prove that the largest number of sightings occurred when Mars was at its closest to Earth. Others, however, argued that the increase in the number of Saucer reports when Mars was in opposition could be attributed to the planet's extreme brilliance at that time. And it was actually distorted views of the red planet that people were seeing! They explained that atmospheric refraction and dispersion can cause a planet to appear to have unusual shapes, flash brilliantly with red, green and blue colours, and seem to move up and down, sideways, and back and forth. All factors, of course, attributed to UFOs.

Despite this very logical explanation, there are still plenty of stories of alleged contact with 'little green men' from Martian Flying Saucers. In 1948, for instance, a spaceship was said to have crash-landed near Mexico City and from it was pulled a twenty-eight-inch high creature complete with boots and loincloth!

Two years later another UFO said to be from Mars crashed at Spitsbergen in Norway. Military experts from both America and Britain were called in to investigate, but their findings were kept tantalizingly classified.

More curious still was the story told by British UFO investigator Cedric Allingham in 1954. For he claimed to have been an eyewitness to the landing of a Flying Saucer near Lossiemouth in North Scotland and then to have made direct contact with its occupant. A curious and rather out-of-focus photograph which Mr Allingham took of this being who said he was from Mars, showed a man six feet tall dressed in a dark, tight-fitting garment. Sceptics who scoffed at the whole tale made a point of ridiculing the pair of *braces* the 'spaceman' appeared to be wearing!

Arguably the most baffling and still unexplained account of all concerns the injured survivor of a UFO which crashed in Poland in 1959. The alien was taken to a hospital where doctors struggled to remove his metallic-looking suit. And as soon as they pulled off an unusual armband, the being died. The body was later transferred to Russia – where interest in UFOs has been just as keen as in the West – and there it was found that the 'Martian' not only had an unusual number of digits but his blood and internal organs were quite different from those of human beings.

Such 'diversions' aside, the idea of life of some kind existing on Mars was again exercising the minds of serious scientists and astronomers in Russia at this time. A prominent figure was the energetic octogenarian Professor Gavril A. Tikhov, the director of the Alma-Ata Observatory in Kazakhstan and a life-long admirer of Tsiolkovsky and the other pioneers. He was also the leader of a special group of Soviet 'astrobotanists' who had set themselves up in the fifties to speculate and write about the possibility of plant life on Mars which could serve as man's nourishment when he finally reached the planet.

Tikhov said he had spent sixty years observing Mars, as well as studying the plants in the Pamir mountains and Arctic areas which he believed equated with those that would be found on the planet. As a result he could now claim 'with sufficient certainty' that there *was* plant life on Mars. He stoutly denied that there was any guesswork in his thesis and said that the red planet had an atmosphere containing moisture and there was precipitation 'fairly regularly' during the Martian year.

The professor claimed the atmosphere of Mars consisted basically of nitrogen, carbon dioxide and a small quantity of oxygen – but he thought it likely this became more plentiful the closer one moved to the surface. 'The "breathing" process of Martian plants is probably facilitated by the oxygen contained in the planet's soil or by the oxygen produced in photosynthesis,' he argued, 'as this occurs for instance in the breathing of marshy or water-dwelling plants on Earth.'

The redoubtable Tikhov further declared that the success of the Sputnik and Lunik missions encouraged him to prophesy that 'not far off is the time when Soviet astronauts will deliver to our Earth the first herbarium of Martian vegetation.'

Other of the professor's 'astrobotanist' colleagues went as far as to say that the Martian plants probably resembled the mosses and lichens found on Earth, as well as our dwarf trees and bushes. A few even thought simple micro-organisms might be found, because on Mars 'there exist the conditions for life development that are the best of all that are available to all the other planets of the solar system, except the Earth.'

All of these men agreed that any human beings landing on Mars would have to wear space suits and live in airtight compartments with rigorously controlled environments. Yet they believed there were no gases injurious to human life, the temperatures would prove endurable, and existence would prove 'considerably easier than on the moon'. Tikhov himself added, 'While on Mars, man can easily do observations, explorations and discoveries of great value to science, in the realm of geology, biology, botany and astronomy. I am sure that useful minerals and metals will be found by men visiting or settling on Mars.'

It was an almost Utopian Mars that Tikhov pictured: a kind of Shangri-la of the solar system. And indeed American scientists also studying Mars at this time had come to a similar, if more reserved, judgement. As Dr John Lederberg, the Nobel Peace Prizewinner, put it in September 1961, 'The most plausible explanation of the astronomical evidence is that Mars *is* a life-bearing planet.'

However, Dr Lederberg and his colleagues were in agreement that though there might be life – it could also present heretofore unconsidered problems.

Chapman Pincher, the *Daily Express* science correspondent, described these in a somewhat dramatic dispatch he filed on 28 September 1961.

'Evidence that life of some kind may exist on other planets is now so strong that a special branch of science called exo-biology has been formed in America to study it. Leading exo-biologists are now deeply concerned about the possible danger of an invasion of the Earth by living organisms from Mars or the other planets.

'What they fear is that robot missiles or manned spaceships returning from these planets may come back contaminated with out-of-this-world bacteria or other miscroscopic organisms.

'If worldly creatures have no built-in resistance to them, these organisms might spread rapidly like a virulent plague infecting people, animals and crops. Cosmonauts will therefore have to go into quarantine for several weeks after returning from another world.' (Interestingly, the exo-biologists' warnings were heeded, and the sterilization of both men and machinery going into space is now, of course, one of the prime requisites of any mission.)

As I showed in the earlier chapter dealing with the American and Russian touchdowns on Mars, we have now established that there is *no* fauna whatsoever of the kind Tikhov and his friends envisaged. And though the Mariner and Viking missions were hard-pressed to reach any definite conclusions from the samples of surface material they *did* examine, these did not altogether rule out the possibility that there might be hidden water deposits beneath the surface of the planet which could possibly contain micro-organisms.

A number of American and Russian scientists now believe that Mars may well have had several episodes of different climates, the last of these – creating the desert-like terrain we now see – having begun a billion to three billion years ago.

The NASA geologist Harold Masursky, who worked closely on the Viking mission, says that the orbiter photographs have revealed a number of channels that could well have been created by the run-off of rainfalls. And he points out that the two Vikings measured very different temperatures as they passed through the planet's upper atmosphere.

'What this means is unclear,' Masursky says, 'but it might

indicate that the forces that govern the Martian atmosphere are delicately balanced. Perhaps Mars experiences brief violent changes in its atmosphere as peculiar to its normal milieu as thunderstorms are in a terrestrial desert.'

Masursky thinks the 'freeze-dried' Mars we now see is its normal self. 'Mars clearly seems to be in an ice age, as is, technically, the Earth,' he says. 'Looking back over the Earth's history, having glacial ice caps at our poles is very unusual. If we have such unusual situations, maybe Mars for periods is a very hospitable place.'

For periods a very hospitable place? Does this imply that though the red planet currently appears to have no visible life forms, they may once have been present in an earlier age? This is a theory which has been gaining ground in both the West and in Russia, and one which I also find particularly fascinating because of the curious and so far inexplicable features which have been observed and photographed on the surface: features which give every indication of having been artificially made rather than created by nature. If so, they hold perhaps the greatest incentive of all for man to reach Mars.

As far as I can tell, the first time that the theory was seriously advanced that there may have once been intelligent beings living on Mars before the atmospheric conditions became too hostile (outside the realms of fiction, of course) was in 1959 by the Soviet astrophysicist, Professor I. S. Shklovsky. The professor, who was in charge of the Shternberg Institute's Laboratory of Radio Astronomy, said that after years of observing the planet he was convinced that 'two or three billion years ago' there were living beings on Mars.

Shklovsky said that he had also made a detailed study of the two Martian moons, Phobos and Deimos, and from the behaviour of their orbits had concluded that they were not composed of solid rock, but were hollow, and therefore might just be artificial satellites. And who else could have built them and put them into orbit other than some long-deceased Martian race? he argued.

Sadly for the doctor's theory, the Viking circumnavigation of the little satellites has established that they are very much natural objects, even if their origin is somewhat in question.

A corollary to Shklovsky's hypothesis was, incidentally, offered by another Russian, Professor G. S. Davydov, who suggested that such an intelligent race, if it ever existed, adapted to a changing atmosphere on the surface of the planet by 'going underground' to find the conditions necessary for their survival. His idea was predicated on the belief that Mars was 'a planet of ebbing life' having once possessed an oxygen-rich atmosphere which leaked away. Since there are no observable lakes or oceans on Mars now, the theory assumed that the surface water froze and was then overlaid with dust. As I mentioned earlier, the polar caps with their icy or frosty coating give some credence to this idea.

Undoubtedly, though, the most persuasive arguments for the idea of a lost civilization on Mars are the strange features pictured by the Viking spacecraft in 1976. Let us look at them and consider the possibilities they throw up.

The hint that the photographs Viking had beamed back to Earth contained anything untoward came not long after the spacecraft had arrived on the planet in July of that year. And the admission was made by the leader of the Imaging Team at NASA headquarters, Tim Mutch.

After studying the pictures he said, 'The Martian landscape is characterized by blocks that litter the surface. The interest stimulated by these blocks depends on your point of view. Viewers lacking in imagination see only a rock pile, a sort of junk heap.

'At the other extreme,' he went on, 'viewers with a surplus of imagination see all sorts of artifacts, even remnants of former civilizations. Somewhere in between, planetary scientists are impressed by block shape, texture and colour – all clues to rock origin and erosional modification.'

A fairly startling idea to be put forward by such a calm and precise man. Yet looking at the pictures one can readily understand the emotions they generate.

The first photograph to excite comment was the picture taken by the Viking 1 lander not long after its touchdown. Going through the thousands of images being flashed back to Earth, one of the NASA scientists suddenly pointed to what seemed to be the letter 'B' etched onto one of the nearby surface bedrocks. Colleagues who clustered round saw the mark just as vividly, though a few argued it might instead be the number '8'. Whatever it was, there was no denying its existence. Apart from the mark, finding bedrock at the landing site also surprised the NASA men. For among other things, it indicated that Martian geology was far more complex than had been predicted, and meant that if a thick layer of ejected deposits once covered this region, it had been subsequently removed by wind, water, or maybe even ice in the form of glaciers. It was another tantalizing step in the quest for signs of Martian life.

Following further study of the picture, the consensus view was that the strange mark was 'an illusion caused by weathering processes and the angle of the sun as it illuminated the scene.' This opinion was not shared by all the scientists, however, and additional research brought to light other rock markings in neighbouring locations. The letters O, S and T were spotted with some regularity and caused one geologist to admit, 'Superficially, they *do* resemble organized symbols.'

But could they be the graffiti of some lost civilization? No one is prepared to confirm *or deny* that just yet!

If the photograph of the letter 'B' was a surprise to the scientists on Earth, it was nothing to the sensation that greeted the picture which Viking took from its orbit of 1,162 miles above the planet's northern latitudes. For while searching for a suitable landing site, it snapped what appeared to be a *huge face* on the surface of the planet!

Curiously, this picture was transmitted to Earth *before* the images of the surface markings, but in the general excitement which understandably accompanied the actual landing of the spacecraft shortly afterwards, it was overlooked for some time.

The photograph was of a seemingly unremarkable stretch of desert in the Cydonia region. But there at 41° north and 9° west was a shadowy face almost one mile across and looking remarkably like one of the famous Egyptian sphinxes on Earth – and just as enigmatic. Closer examination revealed a pyramid arranged symmetrically in what one viewer almost immediately suggested to be the ruins of a city. Were these the remains of the ancient Martian civilization about which Earthmen had dreamed for so long?

The official NASA view poured cold water on such an idea. 'This huge rock formation which resembles a human head,' it said, 'is formed by shadows giving the illusion of eyes, nose and mouth.'

Others, however, felt very differently.

Most notable among these disbelievers are two computer scientists, Vincent DiPietro and Gregory Molenaar, who worked at NASA's Goddard Space Flight Center in Maryland, and another computer specialist, Richard Hoagland of Oakland, California. By applying sophisticated image-processing techniques to the photograph, the two NASA men have been able to reveal details of the left, shadowy side of the face, uncovering an eye, a cheek and a continuation of the mouth, as well as showing what resembles an eyeball with a visible pupil in the eye-socket on the visible side.

'This processing,' says Richard Hoagland, 'effectively eliminates the idea that the face is a trick of the light. It clearly points to it being the result of artificial construction.'

It was while he was examining DiPietro and Molenaar's work that Hoagland discovered what may be the key to the formation's origins and reason for existing. For in studying the surface features sited to the west of the face he came across a gridlike pattern of rectilinear markings like the layout of a city. He also found a series of right angles which gave the overall impression of a main avenue leading towards the face, and a series of walls. Further, he confirmed the existence of the pyramid first spotted by the two NASA men.

Although the computer scientist is reasoned enough to admit that the lines *could* have been by-products of the photo-enhancing process, he is totally convinced of what he saw. For one thing, he says, the objects in the city cast shadows. For another, DiPietro and Molenaar claim they did not get these kinds of glitches with enhancements of aerial photographs taken from Earth. And there is no ready geological explanation for rectilinear patterns of this scale on Mars.

Hoagland claims that if this theory is correct, then the objects were probably constructed 'at least half a million years ago, when according to some current beliefs Mars had a warm, wet period.' He adds that the geology of the Cydonia region suggests that the objects were on the shore of an ancient lake. And the face, over which the sun rose directly, would have formed an island, with the pyramid on the shore beside it.

Though one is eerily conscious of the shade of Percival Lowell hovering over this theory, the grid spacing does resemble that of city streets, and the layout is aligned towards the winter solstice sunrise. And as Arthur Stopes, an architect with a special interest in space matters has pointed out, it makes sense for buildings to have been oriented in a manner that would best use the scant winter sun's warmth.

Therefore, says Richard Hoagland, the conclusion is inescapable: these Martian structures are the product of an intelligence at work. 'Was it coincidence,' he asks, 'that one half million years ago celestial mechanics made the Martian sun rise over a remarkable likeness of man's image?'

Not everyone shares Hoagland's hypothesis about the structures, however, yet he is determinedly pursuing his investigations. And a particular thought spurs him on. 'Now that we have glimpsed what may be waiting in our investigation of Mars,' he says, 'we face the question, "What else lies undiscovered or ignored on the Viking project's remaining one hundred thousand magnetically recorded images?"'

From the work of these three men an even more startling supposition has emerged. Could the objects be the work of alien beings who were not *natives* of the planet?

Let science correspondent Adrian Berry, who has made a special study of these phenomena, explain. 'The theory is,' he says, 'that any beings who walked on the surface long ago and who left the supposed artifacts were travellers from another solar system. They would then have perished on Mars, or departed as they came.

'A still more ominous possibility is that the aliens, if they existed, may have left something potentially deadly on the surface. And if they created the "face" to attract attention, they might also have left a "library", a store of technological information such as would have been amassed by a star-flaring civilization.

'This, of necessity,' Mr Berry adds, 'would be of so advanced a character that it would compare with a description of our own civilization as seen through the eyes of people of the Stone Age. And what a prize that would be!'

What a prize indeed – and perhaps yet another element spurring the race to Mars? Arguably, as I said earlier, even the most *intriguing* one of all. . . .

To summarize, it has to be said that the scientific establishment does not yet subscribe to the idea of ancient civilizations having existed on Mars. But no one now will dismiss the possibility of intelligent life – or at least life – existing on Mars.

For instance, NASA's Mars expert, the engineer James E. Oberg, who has vigorously disputed the theory of a 'metropolis' on Mars, said only a short while ago:

'I do think the most likely evidence for extraterrestrial intelligence will be artifacts we will stumble across and I doubt if they will be small. So searching for artifical structures is legitimate – as long as the facts and arguments are sound.'

And his colleague Harold Masursky, one of the world's leading experts on Mars geology, has added from his most recent deliberations, 'I am sure Mars has had water as well as long periods of severe drought throughout its history. It is farther from the Sun than the Earth is, and its water was trapped as ice for much longer spans of time.

'All of this means that there were fewer eras of favourable climate for intelligent life and civilizations to evolve. But I cannot say there were no civilizations on Mars.'

So the prize awaits. The challenge is on. And the last lap of the race is ready for the winning.

SEVEN

How the Race Will Be Won

AFTER CENTURIES of dreaming, the planet Mars is now undeniably within man's grasp. We need no longer only *imagine* travelling there and landing: we have the equipment and technology actually to do so. It merely requires the initiative from those in authority to put men on a new world for the very first time in the history of our species. And as I have discovered from my research, both America and Russia are within an ace of doing just that, and by way of space missions that are strikingly similar.

Nor is this a possibility we have just attained. As long ago as October 1977 with Viking busy on the surface of Mars, the chief of NASA's Extraterrestrial Materials Research Programme, Bevan M. French, declared with some conviction, 'Even as the Viking data continues to flood in, there are active discussions about follow-up missions to Mars that can now be planned on the basis of what we have already learned. For all that Viking has done, it is only a beginning; what we have learned from the robots on Mars is still not much more than we had learned from the robots (Surveyor spacecraft) that we sent to the moon before the first astronauts landed there.

'We know that the surface of Mars will support the weight of machines and humans,' Mr French went on. 'We have the first rough chemical analyses of the soil. We have taken pictures of the surface and dug trenches in it. And we can now make excellent maps of the planet and pick the sites for future landings. To send astronauts to Mars would be a major undertaking . . . the Vikings have become a bridge into the future. When the landers have sent their last data back to Earth, they will remain like monuments on Mars, waiting silently until new machines, and finally human beings, come to stand beside them.'

A decade on, these words are being echoed with greater emphasis by Thomas O. Paine, the man who was NASA's administrator during the first Apollo moon landing. 'The Viking probes have now transmitted years of environmental observations of the Martian surface,' he says, 'providing us with sufficient design information to construct the first permanently manned Martain bases. And because the Apollo, Skylab, shuttle and Salyut-Soyuz programmes have effectively used most of the components needed for Martian spaceships, the technology to extend manned operations to Mars exists on our drawing-boards today. The availability of water on Mars will provide a favourable climate for the development of agriculture and robotics industry.

'In fact,' Mr Paine adds with the confidence of personal knowledge, 'we know far more about manned operations on Mars today than we knew about lunar landings when the Apollo programme was launched in 1961.'

Mr Paine believes that because settlement on Mars is far more of an essential than it has ever been on the Moon, development resources should not be expended on one-shot Apollo-type manned expeditions.

'Future robotic and manned visits should be designed to prepare the way for permanent settlement and successful colonization,' he says. 'To facilitate such a historic goal, every Mars mission should leave behind materials, sundry supplies and equipment, with qualified men and women remaining on Mars to work between resupply missions. Instead of flying only hardware and tons of materials that can be produced by robots on Mars, we must also send software – the very seeds and blueprints of man's own humanity.'

Although the Russians' space intentions have deliberately never been made anywhere near as clear as the Americans', the Soviet authorities see Mars as just as vital a goal for them to attain. It has been a frequently expressed desire of the men in the Kremlin to 'colonize the solar system' and more than one Western observer has remarked that they 'appear to perceive the broad human appeal of space exploration more clearly than many people in Washington do'.

This factor has been neatly explained by Bruce Murray, a professor of planetary science and a member of the Mariner and Viking teams. 'We must recognize,' he says, 'that the Soviet Union is exploring space on just as large a scale as we are. I think they're doing it for the same basic reason – because they want to, and their people want to lead in exploration. Whatever the motives of the regime, however cynical they might be, it is nevertheless popular for that regime to emphasize Soviet planetary exploration. It's a symbol to the people of Russia that they are an emerging society, that they are leading in the world's activity.'

Though, as I said, the Kremlin likes to play its cards close to its chest, there have been several clear indications of the country's intentions.

For instance, at the beginning of the series of long-duration stays in space achieved by the Salyut spacecraft, the cosmonaut Georgiy Beregovoy wrote that such missions were important because, 'Their successful development is creating the necessary conditions for interplanetary flights.'

This view was supported by another of the cosmonauts, Georgiy Grechko, who put the missions into perspective in 1978 when he commented, 'In less than twenty years the duration of space flight has increased from little more than an hour to many months. This progress will continue, and soon we will be able to stay in space for two or three years. Specialists would not be surprised if men land on Mars in the next twenty years.'

In 1980, the leading Salyut cosmonaut Valeriy Ryuman told Soviet radio listeners that he and his fellow spacemen were more than prepared to undertake increasingly longer missions. 'If an expedition to Mars were being prepared,' he said, 'and it should be necessary to hold a year-long stay in space as an intermediate step, I think that we would readily agree to such work.'

The chief Soviet space doctor, Oleg Gazenko, admitted much the same thing when addressing a conference in Europe in November of that same year. 'It is difficult to give an exact date for a flight to Mars,' he said. 'But I think the basic prerequisites for such a flight exist now. . . . Whether the flight happens in twenty years, I cannot say. But I believe it will be before the year 2000.'

This estimate was revised down still further when the leading Russian space designer and former cosmonaut, Konstantin Feoktistov, spoke to a group of scientists in Moscow in 1984. He said his country could be ready to send a manned mission to Mars in 10 to 15 years, and that such a mission was 'fully within reach' of Soviet technology. He also made a further and even more revealing admission.

'I don't think that a flight to Mars will be carried out earlier than 10 to 15 years from now. But certainly the evidence of life on Mars discovered by unmanned spacecraft *could* spur a Soviet manned mission to the planet. . . .'

James E. Oberg, America's leading expert on Soviet space plans, believes that these forecasts are not idle boasts, but based on sound engineering logic.

'Salyut is just the kind of module needed for a two to three year round trip to Mars in the mid 1990s,' he says. 'Current and near-future orbital operations are generating precisely the kinds of test data needed to design manned interplanetary spacecraft. In fact, the kinds of equipment which will be installed on permanent space stations will not differ significantly from the kinds of equipment needed for a Mars expedition. The life support (with most water, air and food recycled), the crew systems, the communication systems, the navigation computers – all will be ready, after having been used operationally on orbital space flight for years.

'Soviet pronouncements on such distant goals, then, are more than just daydreams; cosmonauts and space engineers are doing their homework now to allow those dreams to be realized. . . . This will all be in fulfilment of the dreams of Tsiolkovsky, which the Russians believe in and have loyally cherished,' he adds.

It is evident, then, that the Soviet authorities see their space stations as paving the way to interplanetary flight. And you can bet that having undoubtedly monitored the remarkable Viking discoveries, they are all the more anxious to go seeking the prize which is on offer. Indeed, it comes as no surprise to me to learn as I write that they are at this very moment preparing a preliminary probe mission to the red planet in 1988. I shall come to this fact again a little later. . . .

□

When I flew to Washington recently to investigate the latest American plans for Mars, it very quickly became evident to me that the impetus to these plans had been brought about by three factors. Firstly, the success and flexibility of the shuttle, now making space flight almost commonplace. Secondly, President Reagan's support of NASA's $8 billion project to create an orbiting space station by the 1990s. And thirdly, of course, the thought that the Russians are planning to push deeper into the solar system.

It became apparent to me, also, that the efforts of a group of enthusiasts who call themselves 'The Mars Underground' were proving similarly important. This was the movement which had kept the pressure on for an American Mars mission during the early 1970s when the dream had suffered a temporary setback. To be sure, between 1960 and 1966, NASA had commissioned 60 studies on the idea of a Mars mission, but this period had been followed by one of austerity at the Washington headquarters (brought about by the budget slashes of the Vietnam era) when such ambitious projects were shelved in order that the then most important of the agency's plans, the shuttle, might be successfully launched.

The Mars Underground had been formed in 1977 to study and expand the idea of a manned mission. Within a few years, and helped by some well-attended conferences, the movement had not only recruited the support and help of scientists, engineers and designers, but also NASA itself.

'Those conferences,' one NASA official told me, 'helped give direction to our efforts. For a Mars mission we need a flexible, evolutionary and technically sound long-range plan – and that's just what such gatherings can help develop.'

At the end of three years, the group, broken up into teams working on spacecraft design, mission strategy and the development of a Mars base, had completed their study which they published as 'The Case for Mars'. They concluded that by using existing technology and equipment a small expedition *could* be successfully mounted costing less than $40 billion – or half the cost of the Apollo moon project!

The document excited all those Americans who yearned after the red planet, and even impressed the hard-nosed bureaucrats in Washington. There it was studied by the Congressional Office of Technology Assessment who – much to the delight of the group – recommended in a report released in September 1984 that Mars should be explored 'with an eye towards settlement as a long-range American goal.'

The scenario for the mission which The Mars Underground presented has now become the basis of American plans.

It will begin on Earth with a booster rocket carrying up into near space the component parts of the Mars ship. In Earth orbit, these parts will be assembled in a Y-shape by astronauts probably using an already operational space station as their base. These cylindrical modules will, in fact, be very similar to those already used to assemble the station, and will be linked with fuel tanks and rocket motors for the 190-million-mile journey to Mars.

Once completed, the Mars ship will be able to blast out of orbit with considerably more ease – as well as with a tremendous saving of fuel – than if it had left the Earth's surface. It will then take course for its six-month coast through space, for all the world looking like some lazily-spinning assemblage dreamed up by Heath Robinson.

Three days before arriving at the red planet, fifteen members of the crew (a party consisting of both men and women) will leave the space ship in groups of five, take control of the small landers integrated into the 'body' of the ship, and prepare for the descent to the planet's surface. These landers will be about half the size of the present space shuttle, and probably wedge-shape in appearance.

Once free of the 'mother ship', the landers will use the thin Martian atmosphere to slow themselves down, as well as parachutes which will save the retro-rockets until the very last stage of touch-down. According to another NASA scientist I talked to about this landing, 'There will be people back home watching this touchdown who will see immediate similarities with the Apollo landings on the moon.'

But that is where the similarities will end. For by no stretch of the imagination can the world on to which those fifteen pioneers descend be described as being anything like the moon. And once having put their footmarks onto the dusty, desert-like surface, planted a metallic flag and begun to observe their new surroundings, they will have had to start to come to terms with the fact that *this* is going to be home, not for days but a good few months.

James French of NASA's Jet Propulsion Laboratory in California, who has already put a lot of work into the strategy of such a mission, explains, 'After their six-month journey, Earth and Mars would no longer be favourably aligned, and an immediate return trip would require too much fuel and too much time – up to two years. So with the next favourable juxtaposition of the two planets being more than 16 months away, the group would need to be prepared for a long stay. And they should use the time to set up a permanent base.'

The Mars plan also allows for the 'mother ship' over a period of 18 months to journey slowly back to Earth where the first replacement crew would board it.

In Washington, I was also interested to learn that there was considerable admiration for a variation of this scheme, put forward by Britain's Dr R. C. Parkinson of the British Interplanetary Society who ventured the opinion the Americans already *had* the wherewithal to reach Mars – and by as soon as 1995!

Dr Parkinson outlined his ideas in a paper entitled 'A Manned Mission To Mars' which he published in the Society's journal, *Space Chronicle*, in October 1981. For the project, he proposed using the Heavy Lift version of the shuttle and the high-energy Orbital Transfer Vehicle (both of which are currently nearing completion) and the already proved Spacelab developed by the European Space Agency, which would serve as living quarters for the five crew members. Because of the particular alignments of the planets in our solar system in 1995, he believed this offered an ideal opportunity for a round trip from Earth to Mars – with a swing-by of Venus – and then home again.

Three vehicles (two orbiters and a lander) would leave

Earth's orbit in November 1994, said Dr Parkinson, using a Heavy Boost Stage for the departure. A smaller stage would allow entry into orbit around Mars.

Three members of the crew would then descend onto the red planet in a cone-shaped landing module (similar to that used for the moon landing), again employing parachutes and on-board engines for touchdown. These pioneer 'Martians' would spend just twenty days on the planet, doing survey and reconnaissance work and some short expeditions with a Mars-rover machine.

The other two crew members in the orbiters need not be idle during this time, the doctor added. They could make an auxiliary mission to Phobos, spending ten days observing the little Martian moon, before linking up with their fellow astronauts for the return leg of the mission via Venus.

'Because this scheme uses existing or developing hardware, with the exception of the lander and Mars-rover which actually rely heavily on Apollo experience, it need cost only four or five times that of the unmanned Viking project,' maintains Dr Parkinson. 'And with the use of Spacelab it could be an international venture between NASA and ESA. Yet its value as a scientific and political achievement would be many times greater, while the "spin-off" in many directions could be incalculable.'

This is certainly a bold suggestion, but I remain of the opinion that the Americans will want to reconnoitre Mars with several more automated probes to find the most ideal landing sites before going for a touchdown with a precious human cargo. Among the probes they have in mind, I believe, are a Polar Orbiter which would provide a complete map of the planet's atmospheric, magnetic, geological and geo-chemical properties; Mars Penetrators, fired from orbiters to bury themselves in the surface material and relay vital information; and also a series of computer-controlled flying machines released from an orbiting 'mother ship' in a protective aeroshell and then freed from this 10 km from the surface to fly observation missions under their own power, provided, most likely, by nuclear or solar electrical engines.

In all probability, NASA would also like to land an automated Mars-rover not unlike the Viking spacecraft, though equipped with tracks, and powered to travel for as much as 100 km over the rocky Martian terrain, observing and recording as it goes. The Russians, too, are known to want to send a version of one of their *Lunokhod* vehicles which surveyed the lunar surface, to similarly explore Mars – though like the Americans they are well aware from bitter experience of the difficulty, and cost, of such a project. Cost will, of course, be a major factor in *any* mission, and the feeling among most space scientists I talked to was that only the Orbiter and Penetrator missions would be absolutely necessary to prepare the way for a manned landing. After them, it would be in the lap of the gods and the skills of the astronauts to cope with anything untoward that might happen in a touchdown.

Once safely landed on the planet, the creation of a base in which the new Martians can live and work is the next priority – and here again a great deal of thought and detailed planning has already taken place.

According to the experts I spoke to, the first beachhead base on Mars will be a very humble one, sited on solid rocky ground, as opposed to sand or unstable avalanche areas near the walls of craters or canyons. A region with mild temperatures and the least wind will also be preferable.

The base will be constructed from the modules originally designed for the space station. They will be buried in a covering of the dusty Martian soil to protect the occupants against the solar and cosmic radiation which beats down on the planet's surface. They will serve as living, working and research quarters.

Nuclear generators will have to be erected close to the base complete with cooling towers. A series of solar panels will supplement this system. Rows of 'greenhouses' will also be erected on a suitable site in which to grow food for the colonists.

On the nearest high ground a weather station will be erected as well as a huge antenna to facilitate surface and space communications. For safety's sake, the area designated for the landing and taking-off of the shuttle craft will be a short distance from the base.

The heart of the base will, of course, be the habitable modules, as space-efficient and comfortable as possible. Large recreation areas will be another priority to provide an environment that is conducive to good morale and psychological health.

Among a number of the scientists who have been studying the problems of establishing a base on Mars, the view has grown that such a base need not necessarily be located in the most scientifically interesting area of the planet. The development of sealed 'rover' vehicles will allow for quite extensive expeditions, and the creation of automated robot craft will enable the colonists to perform any kind of hazardous research without immediate danger to themselves. Equally, pressurized suits developed from those used in space will allow the men and women to move freely around the base.

For really far-ranging exploration, however, travel by air is now being considered the most practical solution, and according to scientist Michael W. Carroll, it can be achieved just as efficiently on Mars as it is on Earth.

'With materials derived from terrestrial ultra-light airplane technology Mars planes will be quite practical,' he says, 'employing small hydrazine engines to propel long-winged craft over ranges of several hundred miles. Since the thin Martian air calls for large areas of wing, aircraft may carry lightweight solar cells on their wings' upper surfaces to power small engines underneath.'

Another top American scientist, Mitchell Klapp, has suggested that a dirigible filled with hydrogen extracted from the Martian water may be the best of all forms of transport on the red planet. Mr Klapp also believes that solar heat would add to the craft's buoyancy in the low-pressure environment.

'The top of the dirigible would be covered with solar cells to power systems inside the small manned gondola,' he explains, 'as well as the large fan engines which would move the craft. These ships would not move quickly, but would be simple to operate and excellent for moving loads in and out of rough places.'

Mr Carroll shares this viewpoint and has added, 'With

these transportation vehicles, researchers, explorers, and colonists alike will have access to nearly any location on the surface of Mars. The new Martians will venture into the volcanic wilderness of Elysium or the magnificent Tharsis bulge. They'll perch on the edge of the great canyons at Noctis Labyrinthus to watch the morning fog float away. They'll see the ochre dunes of Hesperia and stand on the brilliant ice of a polar canyon. These people will be on Mars for practical reasons: business, mining, research. But at these awesome places, they may perhaps feel a deeper, more important reason for being here on another world far from home.'

Two other Mars experts believe that essentially practical objectives are the keynote to the exploration of Mars. Bruce Cordell, an authority on manned Mars exploration, is emphatic that 'the discovery of significant ore deposits on Mars would provide the economic stimulus to explore and establish numerous manned settlements.' James E. Oberg asserts that the natural resources of Mars will have to be turned into air and water in substantial enough quantities to make life possible for future inhabitants.

'The land would first have to be terraformed or made suitable for human life,' he says. 'Though this could cost hundreds of billions of dollars, resulting real estate would be worth at least several trillion dollars, by current standards. It's possible, too, that metals and even diamonds will be found on Mars.'

Should this occur, says Mr Oberg, 'Over the coming centuries, the blood-red planet could be transformed into a gleaming, green-tinted jewel, reflecting the spread of human life across its surface.'

Amidst all the excitement among American experts about the possibility of a landfall on Mars, there are a handful of other men whose eyes are if anything more covetously fixed on the red planet's two moons, Phobos and Deimos. Former astronaut Brian O'Leary now of Science Applications Inc has been making an increasingly more urgent, and not-to-say convincing, case for establishing a survey base on one or other of the two little satellites prior to putting men on Mars.

Mr O'Leary explains, 'In addition to affording an exciting

scientific study of the two asteroid-like bodies, such a base could serve as a weather observation station, a communication centre between exploratory parties, and the means for operating remotely-controlled vehicles over almost the entire globe of Mars.'

Mr O'Leary says the key advantage to such a scheme is that the moons are carbonaceous and believed to consist of as much as 20 per cent water. This could be tapped to provide Mars missions with a 'refuelling station'.

The former astronaut is quick to admit that there are as yet no chemical data to support this, but spectrographic measurements have indicated the presence of water and the mysterious grooves on Phobos offer an intriguing further piece of evidence.

'Steam vents are the only plausible explanation for them,' says O'Leary emphatically. 'The tremendous impact that formed the crater called Stickney might have also provided the heat that boiled the water. And the moon's surface dimples could lie above tunnels that are still blocked with ice.

'So if water is there it could easily be extracted by solar furnaces. And electrical power could then hydrolyse the water into the hydrogen and oxygen needed to fuel a homeward-bound rocket. This would save shipping water and oxygen from Earth, and could cut the costs of a mission by $3 to $5 billion,' he adds.

By O'Leary's calculations, the rocket fuel from Phobos could support visits to and from the Martian surface. Space vehicles could leave Earth with just enough fuel to reach Mars, perform a braking manoeuvre through the planet's thin atmosphere, and then top up their tanks with fuel from Phobos. To confirm all this, he says, an unmanned lander should be sent post-haste to analyse the subsurface material.

'This would be *the* mission for paving the way for going to Mars with men,' says one of Brian O'Leary's keenest supporters, Douglas Jones of the University of Colorado. 'A manned mission could follow within ten years of that robot prospector.'

Curiously, while the Americans are unsure about this idea, their great rivals, the Russians, are not. I got an intimation of

146

this while I was in Washington – but confirmation in London from our leading space expert, Reginald Turnhill, editor of *Jane's Spaceflight Directory*.

'There is growing evidence that the Russians plan to bring a Martian orbiter within a few thousand metres of Phobos in 1986,' he says, 'and then fly in formation with this satellite to obtain high resolution photographic and other data.'

Could this be the galvanizing effort for the Russians' race to Mars? The precursor to a space spectacular in 1992 to mark the 75th anniversary of the Bolshevik Revolution? It would undoubtedly have a stunning prestige effect on world opinion.

As everyone in Western scientific circles is quick to point out, the Russians are very much the 'dark horses' of space research, rarely revealing their true motives. Yet the evidence of increased activity is there, as another important British space commentator, David Whitehouse of *The Times*, observed in August 1984.

'Despite statements to the contrary, there is evidence that Russia *did* want to reach the moon first,' he said, 'but on realizing they could not, denied even the attempt. Cosmonauts have stayed in space for up to seven months, about the time needed for a flight to Mars. Is it possible that Russia has in mind the next great space coup, the planting of the red flag in the red sands of Mars?

'Russia is testing a 14-million-pound thrust booster. With such a rocket, a two-year manned fly-by of Mars and return to Earth cannot be many years away. Only two or three of these rocket launchings would be needed to assemble a spaceship in Earth orbit for a Mars landing.'

In the year or so since this observation, further information that has been received in the West has confirmed this belief.

Firstly, the development of the huge G1 rocket capable of putting almost 400,000 pounds into orbit. This 'superbooster' known as the Saturnski is due soon to have nuclear-powered upper stages similar to the American NERVA programme (Nuclear Engine For Rocket Vehicle Application) which was aborted in 1973. Should it become serviceable shortly – and the signs are that it will be – it would enable the Russians to

147

build their Mars space vehicle with probably greater speed and efficiency than the Americans.

Secondly, the Russians' capability at remaining in space has also increased by leaps and bounds. From the initial few days orbiting the Earth, the record has been pushed up to an amazing 237 days by three men in a Soyuz-7 space station in 1984. The implications to be drawn from the achievement of Commander Leonid Kizim, engineer Vladimir Solovyov and researcher Dr Oleg Atkov were twofold, according to space historian, Ian Graham – each depending upon which camp you favoured.

'The obvious answer,' he wrote recently, 'and the one which many space commentators have suggested, is that Russia is committed to establishing a permanently manned orbiting space station. Soviet space officials have said as much themselves. But opinion in the West is split.

'Others are equally certain that the Soviet Union is going for a manned Mars landing by the end of the century. Both camps cite US Air Force reconnaissance photos of Apollo-class and space shuttle type boosters in preparation and reports of flight tests of unmanned shuttle mock-ups as evidence of their respective cases. Of course, both may be correct. An orbiting station could be used as a staging post, a fuelling and provisioning station or even a spacecraft construction centre on the way from Earth to Mars.'

What *is* beyond dispute is that the Soviets have achieved some remarkable successes in the course of duration flights in space. The cosmonaut Valeriy Ryumin, for instance, has now made two six-month flights in quick succession, travelling 150 million miles – further than the distance to Mars! And in the development of life-support technology they have found a way to carry plants to provide food for far-voyaging cosmonauts. 'Only a Mars mission requires this kind of self-contained system,' a NASA scientist told me with a rueful smile. And, in August 1985, they achieved an even more impressive feat of sending some pregnant rats into space and successfully delivering them of their offspring!

More information is now available on the Russian shuttle, the Cosmos 1614, recently described rather disparagingly in

Sketches of the Soviet space shuttle now under development and expected to be utilized in the Russian drive to Mars. (*Observer*)

Washington as 'Shuttleski' because of its 'disturbingly similar' appearance to the American version. It is, as the reader will no doubt have already appreciated, just one more instance of how similar the elements are in the Russian and American plans for getting men to Mars.

The Soviet shuttle has a lighter lift-off weight than its American counterpart, and can carry a larger payload. Photographs of the prototype taken by a US spy satellite over the Ramenskoye test flight centre near Moscow, reveal the spacecraft to be 109 feet long with a 76-foot wingspan, as against the US version's 122-foot length, and 78-foot wingspan. The leading edge wing sweep is 46 degrees, very much the same as the NASA plane, giving them an almost identical appearance. At the moment, it is anticipated that

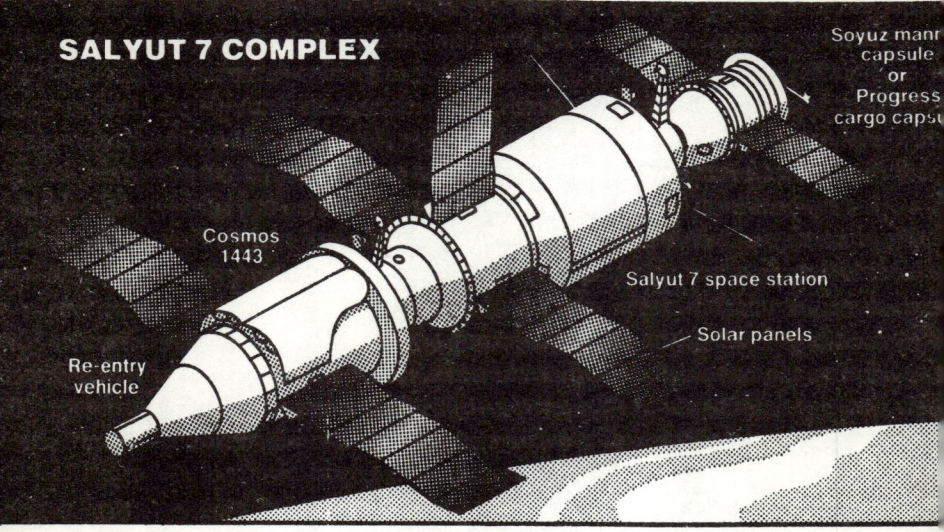

SALYUT 7 COMPLEX

Soyuz manr
capsule
or
Progress
cargo capsu

Cosmos
1443

Salyut 7 space station

Solar panels

Re-entry
vehicle

The Salyut space station complex which will also be an
integral part of the Russians' plans for a manned flight
to Mars. (John Grimwade).

this shuttle will be launched on the D-Type rockets which
have put the Salyut space stations into orbit.

Both the Russians and Americans are fully aware that perhaps the biggest problems of all facing them in any lengthy
space flights are human. The US scientific press has devoted
many column inches to this topic over the years, but not until
recently has it featured in Russian newspapers – another indication, some observers feel, of the Kremlin's usual policy of
overtly alerting its people to possible momentous events.

In August 1985, *Pravda* carried a report about the psychology of long space flights and the 'need to give cosmonauts a
feeling of connection with their lives on Earth'.

For long flights, the paper said, psychologists draw up a
programme of televised meetings between astronauts and
relatives, and home movies of the spaceman's families are
occasionally sent up to the men in the spacecraft by cargo
ships.

'They also listen to tape-recorded sounds of rain, rustling

leaves and the singing of birds,' *Pravda* said. 'It is these sounds, so customary to the ear on Earth, which we may not even notice in the hustle and bustle of daily life, that help astronauts in orbit to overcome fatigue.'

A report in a recent issue of the English journal, *Space Voyager*, enlarges on this fascinating aspect of what lies ahead for those who will journey to the red planet.

'Astronauts must be carefully chosen for their psychological ability to remain stable under difficult and stressful conditions of isolation and confinement,' it says. 'Sufficient scope for recreation and exercise, privacy for individuals, and interesting (not too repetitive) work all help. More problematical are the purely physiological factors of continuous weightlessness, which causes the bones to lose calcium and the heart and muscles to atrophy. The ratio of male to female astronauts is also the subject of study.

'Hopefully,' the magazine adds, 'by the time an actual Mars expedition is planned there will have been plenty of opportunity to solve most of the problems and select the right people by long-term testing in space stations near Earth.'

As I hope I have shown through the pages of my book, this time is nearly upon us now. And certainly behind the closed doors in Moscow and in the corridors of power in Washington, influential forces are at work to try and ensure that one or other of the superpowers is first to have a citizen plant its flag on the dusty red surface of Mars.

Yet this said, I still believe it is not too late to put into operation perhaps the most exciting of *all* Mars projects – a *joint mission* to the red planet.

Already certain leading space figures have cautiously advanced the visionary idea of co-operation between American and Russia to make the first manned mission to Mars an international one. It has even been suggested other nations – including Britain – might join in this plan, so that it is merely members of the human race, rather than any specific nationality, who are first to step onto the new world. A new world which, if this co-operation was developed to its ultimate possibilities, *could* be transformed into an environment where our descendants might live as unconfined as we do today on Earth. And this is no mere idle speculation. . . .

Among the scientists who have floated the idea of co-operation is the much-respected Dr George Mueller, who directed the NASA Mercury programme and the Apollo landings, and first demonstrated formidable powers of accurate prophecy some twenty years ago when he correctly predicted the space shuttle. He was in a similarly forward-looking mood when he addressed the British Interplanetary Society's annual meeting in November 1984 and confirmed to members that human colonization of Mars was 'a practical proposition in the relatively near future.'

Describing how the American and Russian space program-mes were pointing in the same direction with Mars as their objective, he said, 'If we could reach an accommodation with Russia in which we devoted resources we now devote to our military machines to space activities, we could significantly reduce expenditures while carrying out one of the most ambi-tious programmes ever undertaken by man.'

Dr Mueller said he believed the need for nations to co-operate was growing ever more necessary as 'the confines of Earth could eventually destroy man, and human energy should now be applied to developing room for us to live elsewhere'.

'Calculations have shown that a breathable atmosphere *could* be established on the moon, probably by chemical reactions on its surface for several hundred years during which its resources could be developed,' he continued. 'Later on it might be more expeditious to move a moon out of orbit around Jupiter and on to a collision course with Venus or Mars.'

This, the good doctor added, 'would quickly modify the environment of either planet. The excess of heavy gases would diminish on Venus, while the atmospheric gases and supplies of water on Mars would increase.'

This ambitious idea is also believed to be feasible by another NASA researcher who favours co-operation. Maurice Averner foresees the day when human beings will be living on Mars in considerable numbers and will be 'anxious to shed their space suits'. This will call for an ambitious feat of planetary engineering called *terraforming*.

'All that would be needed to give Mars a warm, breathable atmosphere,' Mr Averner says with disarming conviction, 'would be to melt the planet's polar caps, perhaps by focusing sunlight on them with orbiting mirrors. The water vapour produced as the liquid water evaporates would act like the walls of a greenhouse, trapping solar heat to raise the planet's surface temperature. Water vapour molecules would then react with peroxides in the surface to release oxygen.'

And if these reactions yielded enough oxygen, Maurice Averner calculates, then Mars could be habitable 'after only 400 years'. He believes the second stage of this plan – producing a stable ecology of living creatures – would be far more difficult, but as he adds with a smile, 'The risks and rewards are planet-sized!'

Indeed they are. And it pleases me to be able to report that there are a number of English scientists – and laymen – who favour co-operation, one of whom has already devised a scheme to prepare for colonization of Mars by simulating a Martian colony right here in Britain!

The idea comes from a group called Argo Venture, headed by the sociologist Lord Michael Young of Dartington and academic Dr Tony Flowers, and is endorsed by former American astronaut, Russell Schweickart, now the Energy Commissioner of California. They have been pressing the European Space Agency to make a commitment towards the colonization of Mars, and have prepared the blueprint for a mock colony consisting of about 30 men and women from Europe and North America who will reside in an isolated rural or island spot in Britain and there perform the same kind of tasks they might have to do on Mars. They will, though, have computers, videos and other space-age accoutrements.

Lord Young hopes that the simulated colony will help determine the social and political conditions best suited to Mars. 'We want to find the proper balance of sexes; whether there should be children; the right mix of personalities; what authority should be invested in whom and by what right; and what steps should be taken in the beginning to reduce the risk of wars on the new planet,' he says.

Along with the social experiment, Argo Venture will carry out major biological tests to determine whether specially designed bubble chambers simulating the Martian atmosphere can be modified by the introduction of oxygen-producing algae from Antarctica.

'If vegetation grows in these chambers,' explains Lord Young, 'it may mean we'll be able to create a new Martian ecosphere in which the proportion of oxygen on the planet could be raised to sustain human life.'

The adventuresome lord is optimistic that a colony such as the one proposed by Argo Venture could be established on Mars in the early part of the next century. His optimism about future co-operation between the space nations was also raised by a report from Moscow in May 1985 that the Soviet Union and the European Space Agency had 'agreed to co-operate on a new Mars probe to be launched in 1988', according to the authoritative news agency, Tass. It is from such little acorns that big oaks grow. . . .

There are also just the faintest glimmerings of hope that the US and USSR might be coming closer towards co-operation. A possible space mission involving the American shuttle and an orbiting Russian Salyut space station in an experimental 'rescue' of marooned cosmonauts and astronauts has been proposed, a spokesman at NASA told me. This would involve the shuttle rendezvousing with the Salyut, flying in formation, and transferring personnel from the spacecraft to the space station.

'The White House to whom we've talked about this idea seem to think that if the Russians would agree, it could lead on to joint exploration of the moon and even a joint manned flight to Mars,' the NASA man added.

I wait with bated breath for further information – as, I am sure, do the three American astronauts and two Soviet cosmonauts who were reunited in Washington in July 1985 just ten years after their historic meeting in space, and also used the opportunity to urge their governments to undertake a joint mission to Mars.

The Americans were Thomas Stafford, Donald Slayton and Vance Brand who docked their Apollo spacecraft with the Russian Soyuz craft on 17 July 1975 in the so-far only joint mission between the countries, and there shook hands with their Russian counterparts, Aleksei Leonov and Valery Kubasov. Interestingly, it was Leonov who spoke for them all when he told the celebration sponsored by the American Institute of Aeronautics and the Planetary Society, 'I know that all big things start with small steps. But we can accomplish big tasks, not only in space but on the ground as well. I know *we* want to work together.'

In calling for co-operation, Leonov confirmed that the Russians were intending to send a spacecraft to land on Phobos, and that this would take place on 1 May 1989. The Russian's hopes were also echoed by the other four men, and cautiously supplemented by James Beggs, the NASA administrator, who said that experts were agreed that 'with proper political backing' such a mission *could* be under-way as early as 1995, at a cost of about $40 billion.

'But any Mars landing must also include planning for subsequent sustaining operations,' he said. 'And it might be several decades before an undertaking of such scope could be accomplished.'

Although no formal offer has been made, the idea is clearly feasible and the key to its success undoubtedly lies in the development of the relationship between the two superpowers. Without, though, a closer understanding and deeper trust between Washington and Moscow, it seems the race for Mars will go on separately.

Such a state of affairs will particularly sadden Carl Sagan, the President of the Planetary Society, who also attended the symposium. He perhaps more than anyone has been a leading advocate of co-operation, and maintains that a joint US-Russian mission could put human beings on Mars not later than the year 2010. He will not allow his optimism to be daunted, however.

'I know because of the costs it is very hard to justify human missions to Mars solely on the grounds of science,' he says. 'But I can imagine circumstances in which it might be done for other reasons. Suppose the people of Earth are one day fortunate enough to discover new leaders in Washington and

Moscow dedicated to a new beginning; and to seal that new beginning they embark on a dramatic new enterprise – something like the Apollo programme but with co-operation, not competition, the goal?

'Major space missions could also ease the transition in the aerospace industry from the present frenzy of military preparation to more benign activities. Could we muster a mission to Mars with human crews for the sorts of money repeatedly allocated for weapons systems on Earth? Astonishingly, the answer seems to be yes.'

Mr Sagan adds, 'Those rusting Viking landers on Mars represent a symbolic first presence of the human species on another planet. They are reminders of what else is possible for us. The same technology that propels apocalyptic weapons from continent to continents could also provide the first human voyage to another planet. By no means is such a mission to Mars the only – or in my opinion even the best – use of the money we could save if we stepped back from the brink of nuclear annihilation. But it might represent a real choice of fitting mythic proportions: to embrace either the planet named after, or the madness ascribed to, the ancient god of war.'

And to those sentiments I can only add: and may the gods be with us as we make the choice.

Washington – London,
December 1984 – September 1985

156

Postscript

The Summit meeting in November 1985, between President Reagan and the Russian Premier, Mikhail Gorbachev, has further enhanced the idea that a joint USA-USSR space mission to Mars *may* be feasible.

Prior to the actual meeting in Geneva, Mr Gorbachev told a meeting of disarmament campaigners in Moscow that he would 'suggest to President Reagan a joint space mission to Mars'. The Russian leader was no more specific than this in his proposal, but it was evidence, observers believe, of a willingness to work with America towards an achievement that both nations obviously feel is important.

The Summit meeting was, of course, held behind closed doors so we have no precise way of knowing what the two men said to each other on the matter, although there is an indication that the topic did come up, as the following sentence from the joint statement dealing with Space Objectives reveals:

'The relevant agencies in each of the countries are being instructed to develop specific programmes for exchanges. The resulting programmes will be reviewed by the leaders at the next meeting.'

Is it too much to hope that a joint Mars mission *might* be approved the next time the two Superpowers do come together?

December 1985

Acknowledgements

A great many sources, both public and private, were consulted in the writing of this book, and while some of these sources wish to remain anonymous, I would like to record my thanks to the following without whom – to quote that familiar phrase – this book would not have been written. In particular my friends at NASA Headquarters in Washington who were so generous with their time and information. The Russian Press Agency, Novosty, were also forthcoming to a most encouraging degree.

Among the individuals I consulted I should like to thank Ray Bradbury, Arthur C. Clarke, Carl Sagan, Isaac Asimov, Patrick Moore, Albert Parry, James E. Oberg and Frank Winter. Thanks, too, to a couple of splendid researchers in London, Bill Lofts and Jeremy Bentham, as well as the staffs of the British Museum Newspaper Library, The London Library and The Library of Congress in Washington. And not forgetting these newspapers and magazines who opened their files to me: *The Times, Daily Mail, Daily Telegraph, The Observer, New York Times, Nature, Astronomy* and *Aviation Week & Space Technology*.

Finally, my thanks to my editor, Joe O'Reilly, and also my wife, Philippa, for her usual patience and understanding while I was working against the clock to complete the manuscript before the race was actually won!